ALGEBRA

THE BEAUTIFUL

ALSO BY G. ARNELL WILLIAMS

How Math Works: A Guide to Grade School Arithmetic

for Parents and Teachers

ALGEBRA
THE BEAUTIFUL

An Ode to Math's Least-Loved Subject

G. Arnell Williams

BASIC BOOKS

NEW YORK

Basic Books
Hachette Book Group
1290 Avenue of the Americas, New York, NY 10104
www.basicbooks.com

Printed in the United States of America

First Edition: August 2022

Published by Basic Books, an imprint of Perseus Books, LLC, a subsidiary of Hachette Book Group, Inc. The Basic Books name and logo is a trademark of the Hachette Book Group.

The Hachette Speakers Bureau provides a wide range of authors for speaking events. To find out more, go to www.hachettespeakersbureau.com or call (866) 376-6591.

The publisher is not responsible for websites (or their content) that are not owned by the publisher.

Library of Congress Cataloging-in-Publication Data
Names: Williams, G. Arnell, 1964– author.
Title: Algebra the beautiful : an ode to math's least-loved subject / G. Arnell Williams.
Description: First edition. | New York : Basic Books, 2022. | Includes bibliographical references and index.
Identifiers: LCCN 2021054415 | ISBN 9781541600683 (hardcover) | ISBN 9781541600706 (ebook)
Subjects: LCSH: Algebra. | Mathematics.
Classification: LCC QA150 .W55 2022 | DDC 512—dc23/eng20220517
LC record available at https://lccn.loc.gov/2021054415

ISBNs: 9781541600683 (hardcover), 9781541600706 (ebook)

LSC-C

Printing 1, 2022

IN DEDICATION

To my mother, Geneva Williams, and grandmother, Daisy Bell Rowland, who through their wisdom, foresight, strength, and work ethic carved out stable, safe, and creative environments that made possible so many of the positive things in my life.

IN MEMORIAM

Of Judy Palier and Sumant Krishnaswamy, both of whom made categorical differences in the lives of so many students and others in Northwestern New Mexico and Southwestern Colorado—their legacy continues to thrive there. And of little Charlie Sullivan (Char-Bar), who, though here a short time, demonstrated an inquisitive nature and young sense of humor that brought joy to all who knew him.

IN CELEBRATION

Of all who have attempted to broadcast the power, beauty, and dramatic landscape of algebra.

Contents

Acknowledgments

Writing a book on elementary algebra presented unique and different challenges from writing my first book on arithmetic. The place of arithmetic in the curriculum is rarely questioned even by those who may have struggled with it in school. The same cannot be said of algebra. Algebra in education holds a much more fragile position in the collective psyche of most Americans.

Attempting to improve the perception of the subject in the minds of my students is where I first confronted this challenge head on many years ago. And it is to that large, diverse body of individuals that I owe the greatest thanks for the existence of this book. Their skepticism, intelligence, emotional and conceptual struggles, curiosity, insights, humor, suggestions, and enthusiasm are what have helped to inform and sharpen the techniques and strategies that I have employed over the years.

It is my hope that this book conveys and extends to the printed word some of the excitement and energy from those classroom interactions.

As with any project of this size, the support of others at some point during the process through the reading of drafts, encouragement, advice, and so on has been of the greatest assistance.

Firstly, I would like to warmly thank those who enthusiastically offered their time to read nearly the entire manuscript as it stood at various times. The individuals who gave the full measure were Terri Butler, Mark Pfetzer, Jim Phillips, Brandi Bushman, and Callie Vanderbilt. Their comments from the vantage point of a complete overview of the project were invaluable.

Individuals who read portions of the book and provided valuable commentary include Carl Bickford, John Burris, Carrie Elledge, Mary Fischer, Traci Hales-Vass, Sumant Krishnaswamy, Vonda Rabuck, Jenia Walter, and Jeff Wood.

Many thanks to Bill Hatch for his generosity in providing illuminating artwork in Chapters 8, 9, and 11.

Other friends and family who offered advice, support, and encouragement include Shelley Amator, Jim Barnes, Eric Bateman, Angela Bishop, Rachel Black, Amy Jo Bramlett, Sherri Cummins, Anne Cunningham, Mabel Gonzalez, Jane Green, Susan Grimes, Ken Heil, Vicki Holmsten, Ewa Krakowska, Kris Kraly, Karen Kramer, Lynn Lane, Michelle Meeks, Eric Miller, Jon Oberlander, Alicia O'Brien, Liam O'Brien, Elizabeth Phelps, Gabriela Rivero, Alicia Skipper, Danielle Sullivan, Vernon Willie, and my sister Jennifer Fagbemi.

Much appreciation goes to my coffee shop compatriots: Charlie Travis, Guy Dykes, Louis Eberharter, Susan Girton, David Griffin, Bernard Bro, Cindy Dunnahoo, Lorenzo Brown, Brian Williams, Ed Marquez, Trish Marquez, and the young, energetic, friendly baristas. Their conversation and humor always provided well-needed relief throughout the hours of writing, and I learned a lot from them as well.

Special thanks to Karen Badcock, Carol Jonas-Morrison, Laurie Gruel, and Allan Nass for their friendship, support through my efforts at finding a publisher, spot-on editing of my various communications, and overall wisdom. I can't thank them enough.

A note of particular thanks to David Mumford, who, in addition to taking the time to skim through several of the chapters in the earlier phase of this project (making some helpful suggestions along the way), was also supportive and forthright in his communications with me starting from the time he received a complimentary copy of my first book on up through the present.

Thanks to Michael Schrage for his kind correspondence and words of encouragement on the idea of the book. And to Ian Stewart as well for encouraging me to push forward in my efforts to get this second project published.

Much appreciation to the Textbook and Academic Authors Association for their incredibly informative and friendly conferences that put me in contact with some truly remarkable people. Particular thanks to John Bond for his amazing support and enthusiasm throughout the final years of this project. In addition to his knowledge, his generosity still stands out the most to me. A debt of gratitude is owed to Stephen

Gillen for looking over my contract and helping me to understand it down to the last detail.

Great recognition and thanks to my editor T. J. Kelleher, who showed tremendous confidence in this project and me from my initial query right on through the production process. I appreciate it more than I can express. Special thanks to Marissa Koors for her collaborative spirit, enthusiasm, and extremely productive line-edits. It was an enjoyable and challenging intellectual exercise considering and incorporating many of her ideas on how to improve the book. Much gratitude to Charlotte Byrnes for her extremely thorough and detailed copy-editing of the many faces of the book and the comfort this attention to the particulars gave me moving into the final stages of producing the book. Sincerest appreciation to my production editor, Melissa Veronesi, for her creativity, responsiveness, and attention to detail in overseeing the entire production process. Other marvelous staff at Basic Books owed thanks include Amy Boggs, Madeline Lee, Melissa Raymond, Rebecca Lown, Ivan Lett, Jessica Breen, and other production and marketing staff.

And finally, an incredible thanks to my late friend Deb Mullen, who in my early days of teaching enthusiastically supported and encouraged me to feed the passion of making mathematical ideas clearer to those students who truly wanted to better understand and who needed it the most. I wish I could tell you how special and valuable those inspirational conversations have been throughout my career—to both my teaching and my writing.

G. Arnell Williams
San Juan County, New Mexico
May 2022

People remember their teachers—the ones that excite them to learn. It is what influences and impresses one. Take that to school with you the first day and try like crazy to keep it.

—Deborah J. Mullen (1963–2015),
Naugatuck, CT

Introduction

People rarely come to algebra with neutral attitudes. For many, difficulty with the subject in school defines and dominates their mathematical experience, and they often are quite passionate and expressive about it.

Nor are they historically alone in this feeling either. This intensity of emotion is no recent thing.

In 1749, Frederick the Great of Prussia wrote to the famed man of letters, Voltaire, regarding algebra, "…But to tell you the truth I see nothing but a scientific extravagance in all these calculations. That which is neither useful nor agreeable is worthless. As for useful things, they have all been discovered; and as to those which are agreeable, I hope that good taste will not admit algebra among them."[1]

More recently, in 1930, nationally syndicated columnist Dr. Arthur Dean wrote, "If there is a heaven for school subjects, algebra will never go there. It is the one subject in the curriculum that has kept children from finishing high school, from developing their special interests and from enjoying much of their home study work."[2]

One can find similar sentiments expressed throughout the last four centuries, from the time symbolic algebra was first taught in schools up to this very day.[3]

So, what then is this thing called algebra? What is it truly about? What are some of the things that distinguish it from arithmetic? And what advantages, if any, does it really offer to those who know how to use it?

These are questions that *Algebra the Beautiful* aims to shine a discerning spotlight on, tackling them head on from the jump and continuously throughout the book. Going into great creative detail in explaining some of the basic procedures of the subject (such as why do we use letters of the alphabet to describe unknown and variable quantities), it aims to bring back to life the aura and excitement of their discovery and early use while simultaneously explaining in clear, understandable language why they work and some of the capabilities and enhanced perspectives that they can still give to us today.

Algebra has been in its current symbolic form only since the 1600s CE, but algebraic documentation dates back to the Mesopotamian and Egyptian eras (ca. 1700s BCE) and perhaps even into the Indus River Valley civilization—if we could only decipher their writing. Chinese and Greek documentation has also been noted during the first millennium BCE.

Why did it take so long for smart people to make the conceptual leap to the sleek symbolic representations that we see today?

Algebraic historians still are trying to gain a better appreciation of this, but the gap in time clearly suggests that the symbolic representations in use today may not be as intuitive as some think they are, bypassing many important purposes and details. And therein lies much of their amazing efficiency; yet therein simultaneously lies great difficulty, too, for when we teach symbolic algebra to students, it is easy to make a great many assumptions about their understanding of what the symbols and procedures are actually accomplishing. Assumptions that students, who struggle with algebra, are often tripped up by, especially if they are never explicitly pointed out.

One of the central aims of *Algebra the Beautiful* is to try to get out in front of those hidden purposes and details, making them far more transparent for readers. It does this by focusing on a few accessible topics in great depth and variety. In pursuing this route, the book takes full advantage of the opportunities that open up for exploring nuanced and important matters such as parameters, how the mixing of variation and stability can be gauged through algebraic representation, and the incredible reach and unifying power of algebraic expressions and equations.

The great educational philosopher John Dewey's thoughts, from his 1934 book *Art as Experience*, have been described thus:

An experience occurs when a work is finished in a satisfactory way, a problem solved, a game is played through, a conversation is rounded out, and fulfillment and consummation conclude the experience. In an experience, every successive part flows freely. An experience has a unity and episodes fuse into a unity, as in a work of art. The experience may have been something of great or just slight importance.[4]

Algebra the Beautiful is a math book that seeks to give you many such experiences—experiences that will hopefully transform, for the better, your entire view of the subject of algebra.

It can be likened to tourists experiencing a national park. If the roads are appropriately placed with adequate turnouts, hiking trails, and interpretive centers, visitors can gain a spectacular appreciation of the dramatic scenery, say of the Grand Canyon, without exhaustively visiting every inch of the park.

Algebra the Beautiful aims to make a definitive and lasting impression by showing that algebra uncloaked is big, varied, dramatic, and relevant, forming an interactive, reliable foundation for all of mathematics and many other areas as well. It communicates this to readers through several distinctive approaches, including the following:

- **The Humanistic Approach:** Algebra (as well as the rest of mathematics) is not an isolated island but rather shares similarities with other great areas of human activity, expression, and ambition—including science, language, history, art, music, and philosophy—that seek to better understand and describe the world and then use this knowledge in impactful ways.[5]
- **The Aesthetic Approach:** Mathematicians and scientists frequently state that mathematics is beautiful, yet most nonexperts don't see it that way. In *Algebra the Beautiful*, the aesthetic is interwoven into the very fabric of the book. This is done by reimagining elementary algebra as a vehicle for illuminating the general beauty of mathematics. Ideas work together in concert, and paying nuanced attention to the beauty produced by their interaction serves as a powerful weapon of exposition for the book.

- **The Conceptual Approach: Metaphors, Narrative, and History:**
  Algebra the Beautiful demystifies the techniques of elementary algebra by using metaphors, analogies, and history in unique and robust ways to tell the subject's powerful and holistic story. There is magic in the combination.

Algebra the Beautiful doesn't seek to dazzle you with a stunning display of facts, nor present you with a long list of mathematical formulas. Rather, the goal of this book is to strike at the heart of your conceptual and emotional understanding of algebra, to put you on more intimate terms with a few of the simple, yet elegant, ideas at the core of the subject while at the same time taking you on an imaginative intellectual journey through mathematics itself. In short, this book aims to inform, bolster, and inspire your mathematical soul.

Variables and Motions

Algebraic letters are pure symbols; we see numerical relationships not in them, but through them; they have the highest "transparency" that language can attain.

—Susanne Katherina Langer (1895–1985),
Philosophy in a New Key: A Study in the Symbolism of Reason, Rite, and Art

1

Numerical Symphonies

Music is the electrical soil in which the mind thrives,
thinks and invents.
—Ludwig van Beethoven (1770–1827), letter from
Bettine von Arnim to Johann Wolfgang von Goethe

Algebra is a vast and beautiful continent—at times serene and familiar, at other times mysterious and wild.

Despite the fact that it has been powerfully used for centuries, underwriting some of humanity's most important innovations, serious questions and riddles remain: especially in regard to its essential nature, its place in education, and why so many intelligent people struggle to understand it.

Consider this an invitation to experience some of the vastness, aura, and beauty of this terrain.

But algebra does not reveal its scenery for free. It requires an intense cocktail of conceptual techniques to bring this beauty into sharp relief. Consequently, I will heavily employ some of the most powerful weapons of exposition available, including metaphor, analogy, history, and narrative.

Of metaphor, mathematics education researcher Anna Sfard states:

Metaphors are the most primitive, most elusive, and yet amazingly informative objects of analysis. Their special power stems from the fact that they often cross the borders between the spontaneous and the scientific, between the intuitive and the formal. Conveyed through language from one domain to another, they enable conceptual osmosis between every day and scientific discourses, letting our primary intuition shape scientific ideas and the formal conceptions feed back into the intuition.[1]

Of history, mathematician J. W. L. Glaisher said, "I am sure that no subject loses more than mathematics by any attempt to dissociate it from its history."[2]

And of narrative, cognitive psychologist Steven Pinker says, "Cognitive psychology has shown that the mind best understands facts when they are woven into a conceptual fabric, such as a narrative, mental map, or intuitive theory. Disconnected facts in the mind are like unlinked pages on the Web: They might as well not exist."[3]

Many experts share these sentiments.

In this book, we explore how far we can go with injecting these techniques (with a vengeance) throughout the discussion. My hope is that it will transform your conceptual and emotional understanding of this oft-maligned subject. In this chapter, we begin with music.

MUSIC

Music is one of the most remarkable of all the activities of humankind. Millions willingly subject themselves to its mood-altering effects day after day. Just a simple thirty-second ditty or jingle can launch back to life memories from decades past.

Mute the sound to a video of people vigorously dancing and their energized behavior looks fascinating at best, bizarre at worst. Go to the mall, ride an elevator, or watch a movie, and you will find it there lurking in the background. It is everywhere.

But what exactly is it? Why does it impact people in the ways that it does? How can it launch some into a state of almost pure euphoria while reawakening painful emotions, long thought extinct, in others?

It too is a vast expanse of familiarity, serenity, and mystery.

Of all its forms and manifestations, one of the most grand, vivid, and complex arises in the guise of the symphony: "a lengthy form of musical composition for orchestra, normally consisting of several large sections or movements…"[4]

The trajectory of sounds in a symphony can be extensive, wide-ranging, and dramatic. Reaching a profound and notably intense form in the work of Ludwig van Beethoven, it is said to be the medium in which many composers still choose to demonstrate their technical prowess and most expressive ambitions.[5]

Our interest with it here lies primarily in the great variety and scale that can arise around a central well-developed theme (or core)—we will see something similar happen repeatedly in a mathematical context.

One of the most fascinating things about music is that it is possible to capture the dynamic range of a symphony on flat sheets of paper. It is almost as if musicians can freeze the essence of an hour's worth of lively music and hold it in suspended animation to be viewed later and analyzed at their leisure. This is no small thing and offers great benefits to those who choose to use it.

Written musical notation gave Beethoven (who could barely hear at all by his forties) the inspirational capacity to compose and share wonderful music right on up to the last years of his life, with the release of his highly acclaimed *Ninth Symphony* occurring in 1824 at age 53.[6] It is hard to imagine him doing this without the aid of visual notation. To this very day, orchestras are still able to perform these masterpieces thanks, in large part, to their preservation in written form.

But the sounds of music in a complex performance are not the only phenomena that can vary in our world.

Variations in temperature, moisture, and so on connect up to collectively form the climate of a region.

Variations in events, political leaders, ideas, cultural norms, and so on combine to form the history of a place.

Artificial satellites soar through space constantly changing their individual locations, which, when taken together, collectively form an orbit; whereas weekly variations in time on the job join up to give the yearly earnings of an hourly employee.

Variations in nature differ in kind, too—with some variations being extremely simple (capable of complete description), some being more difficult to tame but still forecastable, and others seeming to totally defy prediction.

A central goal of this book is to learn more about variations of the numerical persuasion and to showcase their accompanying descriptions in symbols. We will find these variations to be relevant, often surprising, and more around us than we might think (often unrecognized). Moreover, we will find that their systematic description opens wide to us an entirely new and vast-reaching branch of mathematics: one that is distinct and separate from elementary arithmetic on the one hand yet critically fused at the hip with it on the other. Together these two branches will team up to form one of the most potent one-two punches in the history of human thought—creating, in the process, a quantitative version of Beethoven's "electrical soil" in which the sibling spirits of mathematics and science can often materialize in, thrive, and discover masterful expression.

MAGICAL THREE-DIGIT NUMBERS

We now take a look at an artificially created numerical variation and observe how the values it produces can store wide-ranging information. To get a good feel for it requires your participation.

Pick the number of days you like to eat out in a week (choose from 1, 2, 3, 4, 5, 6, 7). Multiply this number by 4. Then add 17. Multiply that result by 25. Next add the number of calendar years it is past 2013 (e.g., if the year is 2016, then add 3). Now if you haven't had a birthday this year, then add 1587, but if you have had a birthday this year, then add 1588. Finally, subtract the year that you were born from this.

After all is said and done, you should have a very personal three-digit number. Reading it from left to right, the first digit is the number of times you like to eat out in a week and the last two digits are your age. For those younger than 10 years of age, it also works if their age is simply interpreted as a two-digit number with a zero in front (i.e., interpreting 09 as 9 and so on). Try it again, using a different number of days and/or a different date and year. Save your efforts for we shall return to them in later chapters.

Here are a couple of examples of this in action:

1. Let's say that the current date is November 5, 2030, that Abu Ka-mil's birthdate is February 6, 1950, and that he likes to eat out five times a week. This scenario reads for him as follows:
    a. Pick the number of days you like to eat out in a week: His number is **5**.
    b. Multiply this by 4: His number is now 4 × 5 = **20**.
    c. Add 17: He now has 17 + 20 = **37**.
    d. Multiply that result by 25: This gives him 25 × 37 = **925**.
    e. Add the number of calendar years it is past 2013: In 2030 this would be 17, which would give him 925 + 17 = **942**.
    f. If you haven't had a birthday this year, then add 1587. If you have had a birthday this year, then add 1588: He had a birth-day, so he adds 1588, which gives him 1588 + 942 = **2530**.
    g. Subtract the year that you were born from this: Since he was born in 1950, this gives him 2530 − 1950 = **580**.
    h. Reading from left to right, the first digit is 5 (the number of times he likes to eat out a week) and the last two digits are 80 (his age at the current date).

2. Let's say the current date is May 16, 2018, that Pandrosion's birth-date is December 26, 1980, and that she likes to eat out twice a week. Then this scenario reads for her as follows:
    a. Pick the number of days you like to eat out in a week: Her number is **2**.
    b. Multiply this by 4: Her number is now 4 × 2 = **8**.
    c. Add 17: She now has 17 + 8 = **25**.
    d. Multiply that result by 25: This gives her 25 × 25 = **625**.
    e. Add the number of calendar years it is past 2013: In 2018 this would be 5, which would give her 625 + 5 = **630**.
    f. If you haven't had a birthday this year, then add 1587. If you have had a birthday this year, then add 1588: She has not had a birthday, so she adds 1587, which gives her 1587 + 630 = **2217**.

g. Subtract the year that you were born from this: Since she was born in 1980, this gives her 2217 – 1980 = **237**.

h. Reading from left to right, the first digit is 2 (the number of times she likes to eat out a week) and the last two digits are 37 (her age at the current date).

There are hundreds of different values that can be generated by this process by varying the number of days, the current date, and the ages of the participating readers. Some such numbers include 123, 457, 720, 485, 323, 389, 717, 649, and 234, as well as possible different ones obtained by you and other readers—it is a veritable "symphony" of numbers!

However, the final scenario involving this three-digit number (and our interpretation of it) will fail for someone who is 100 years of age or older.

What in the world is happening? Why do these varying three-digit numbers simultaneously contain information that is personal to each of you yet different from other readers? Is it possible to describe this and to also show why it fails for centenarians? If so, how do we do it?

ARITHMETIC IS NOT ENOUGH

On its own, arithmetic will encounter great difficulty in conveniently describing what is happening in this number of days and age problem. There is simply too much going on—too much variety.

In generating each number, it is almost as if we are doing the same type of arithmetic process but on a different channel (identified by the number of days we like to eat out, the current year, our year of birth, and whether or not we have had a birthday). In the two examples we worked through above, one channel generates the number 237 while the other channel gives 580.

There are many other channels that generate all of the other values that can be produced.

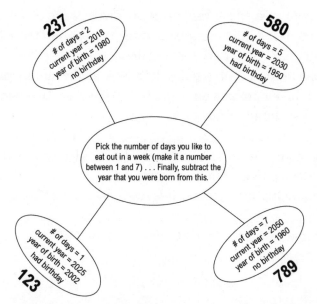

Four channels of the hundreds that can be run and the resulting
numbers generated

This then is a first example of the great variety (of numbers in this case) that can arise around a central theme or process. We don't have enough tools yet to tackle this problem directly, so we will scale back to simpler scenarios and cut our teeth there first.

ORDINARY LANGUAGE IS NOT ENOUGH

Let's begin by considering a different way to express a well-known and basic property of addition: the value we obtain when adding any two numbers doesn't change if we reverse the order in which we add the two quantities. For example, the value we obtain when we add 3 + 5 is 8, and we obtain the same value if we reverse the order of the two numbers and add 5 + 3. This property is more formally known as the commutative property of addition.

You might ask, isn't it enough to simply describe the property as we have above; what's the point of searching for another way? The point being that, while describing mathematical concepts in plain English can be useful for representing and communicating ideas, it is not very useful for systematically rearranging them. Sometimes the key to grasping an idea or concept critically involves the ability to conveniently maneuver it into a simpler or more transparent form. Language simply is not always up to the task of doing this.

As an illustration of this point, let's look at the simple problem of adding the three numbers one hundred sixty-seven, two hundred seventeen, and six hundred eighty-nine:

> Written out this reads as: one hundred sixty-seven plus two hundred seventeen plus six hundred eighty-nine.
> Using mathematical symbols this reads as: $167 + 217 + 689$.

The mathematical form yields to easy manipulations (once we know the rules):

$$
\begin{array}{r}
6\;8\;9 \\
+\;\;2\;1\;7 \\
\hline
1\;6\;7 \\
\end{array}
\qquad \text{becomes} \qquad
\begin{array}{r}
1\;2 \\
6\;8\;9 \\
+\;\;2\;1\;7 \\
1\;6\;7 \\
\hline
1\;0\;7\;3 \\
\end{array}
$$

Conversely, the addition in English words alone does not yield to simple manipulations. That is, there is no realistic method to work our way to the answer using only what we are initially given—namely, the letters of the alphabet:

starting from
  six hundred eighty-nine
  two hundred seventeen
  + one hundred sixty-seven

and manipulating letters only (no numerals) to obtain

  six hundred eighty-nine
  two hundred seventeen
  + one hundred sixty-seven
  one thousand seventy-three

In practice, whenever we are given larger numbers in words to add, most of us resort to using numerals (whether on paper, mentally, or using a calculator) to complete the computation. We don't think to compute by aligning the words and adding individual letters. That is, we don't ask ourselves what adding the last letters of each number word (e + n + n) will equal and so forth—letters when used as language components simply don't work that way.

In a similar fashion, looking at the number of days and age problem as it currently reads doesn't give a clear idea of what is going on. We can certainly run the numbers as the procedure asks, but why they end up the way they do seems almost like magic.

We need a different way to express the problem. We need a method that transforms the problem the way the symbols "167 + 217 + 689" transform the English statement "one hundred sixty-seven plus two hundred seventeen plus six hundred eighty-nine." In short, we need to take the entire problem itself, as stated in English, and recast its essence in a new form—into something that can be operated on and meaningfully rearranged.

In the rest of this chapter, we will focus on how to recast quantitative ideas and procedures that can vary or change value into a more malleable form, then in the next chapter we will turn our attention to how to successfully maneuver them after they have been converted to this new form.

STORING IDEAS

Let's return to the commutative property of addition. We give four more examples of this property in action:

$$\underbrace{1+2=2+1}_{\text{I}}; \underbrace{8+12=12+8}_{\text{II}}; \underbrace{452+987=987+452}_{\text{III}};$$
$$\underbrace{11200+876543=876543+11200}_{\text{IV}}.$$

The total possible occurrences of this property are infinite. And once more we are faced with a "symphony of numbers": this time involving sets of numerical expressions (such as I, II, III, and IV) as opposed to single numbers.

Lots of variety indeed, yet all of it tied together by a simple core theme—that of commutativity (i.e., the order in which we add two numbers gives us the same answer). Like the number of days and age problem, it is as if all of these different expressions are simply different channels of the same idea.

From here on out, when we refer to a "symphony" or "ensemble" it will mean a collection of numbers, expressions, or objects that are tied together around a specific procedure, rule, or theme.

Let's now give chase to recasting this idea of commutativity into a more malleable shape by capturing its operational essence, which is that we have two slots, on the left-hand side of the equals sign, into which two numbers can be inserted and added, and then we reverse their positions on the right-hand side of the equals sign. We can describe this as

first number + second number = second number + first number.

An advantage of expressing the idea of commutativity this way (as opposed to writing out "that the order in which we add two numbers doesn't matter") is that the arrangement now has the same form as the property does when we write it out with numbers (e.g., $1 + 2 = 2 + 1$). That is, the expression now is not very far removed operationally from the thing it is describing (this is not true of the written statement).

In the four numerical examples I, II, III, and IV, the slot described by "first number" takes on the values 1, 8, 452, and 11200, and the slot described by "second number" takes on the values 2, 12, 987, and 876543, respectively. Thus, this expression operationalizes, in a sense, the general idea of commutativity.

If we were text-messaging this idea to someone, we might choose to abbreviate it in either of the following ways:

first number + second number = second number + first number

becomes

$$fn + sn = sn + fn,$$

or taking this even further to

$$f + s = s + f,$$

with no loss of essential information. If we choose the latter, all of the variety that can be expressed with the different numerical instances of commutativity can be reproduced from this stripped-down alphabetic rendering in the following way:

Set f to:	Set s to:	f + s = s + f
1	2	$1 + 2 = 2 + 1$
8	12	$8 + 12 = 12 + 8$
452	987	$452 + 987 = 987 + 452$
11200	876543	$11200 + 876543 = 876543 + 11200$

If we let f = 300 and s = 987, then f + s = s + f becomes 300 + 987 = 987 + 300 and so on. The innumerable demonstrations (variations) of the commutative property in action can now all be obtained by simply setting "f" and "s" to the required numerical values in the expression f + s = s + f.

All of the infinite variation (hence the idea of commutativity itself) is now, in effect, captured and becomes stored or seeded in a single easy-to-read expression. This is a major conceptual shift as we are now looking at letters as platforms for storing changing numerical values as opposed to their traditional use as carriers of information for the spoken or written word.

There is great value in this. We store ideas in language, too.

In English, the word *tree* applies to trillions of distinct and varied plants on Earth.[7] Each individual tree is a tangible example of the specific combination of qualities that we give to the word t-r-e-e: meaning that the word serves as a symbolic storage device for every one of these plants. They have properties in common that allow us to quickly refer to the majority of them as *trees*.

We gain tremendous advantages by being able to refer to lots of different and distinct things by the same expression or name. In this particular case, we can in one statement ("Trees help to remove carbon dioxide from Earth's atmosphere") communicate something that applies equally well to processes involving every single living tree on the planet. A single innocent sentence, simple enough to be taken in with a single sweep, is still broad enough to say, all at once, something that is true about trillions of different plants.

Languages, in general, give us this wide-ranging ability to describe lots of objects and ideas with a relatively small glossary of words. Taking these words, then, in combination to form sentences—language expressions—gives us the breathtaking ability to describe nearly everything that we experience in life or are able to think about in the world around us. We seek the same in the world of numerical variations.

OTHER ENSEMBLES OF NUMBERS

There are other numerical ensembles out there awaiting description. Let's look at a few.

Consider a plane flying at a height of 35,000 feet that is traveling west for four hours at the constant speed of 450 miles per hour. During this time, it travels a total distance of 1800 miles, meaning that, in theory, every distance from 0 to 1800 miles is covered at some point over the four-hour journey.

Now this might be starting to sound like the word problems you remember, and maybe dreaded, from algebra class. However, based on the principles we've just established, this is nothing more than another kind of numerical symphony. Let me show you what I mean by including a few members of this ensemble: 450 miles (for the distance traveled in 1 hour), 900 miles (for the distance traveled in 2 hours), 1080 miles (for the distance traveled in 2.4 hours), 1350 miles (for the distance traveled in 3 hours), and 1800 miles (for the distance traveled in 4 hours).

We can visually represent this:

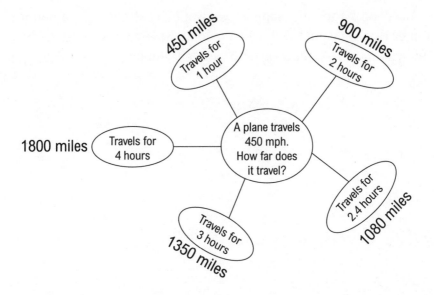

A nice way to store and operationalize all of this variety is by again identifying its core, which in each case here involves multiplying 450 by the time of travel in hours. Here, we will abbreviate "the time of travel" with the letter "t" (for time), and then we can reproduce all of the numbers in the previous diagram by simply writing

$$450t$$

(which means 450 multiplied by t) and setting t to the following values:

$$t = 1; t = 2; t = 2.4; t = 3; t = 4.$$

The expression 450t stores in a leaner form the variation contained in this situation. We have shown only five distances produced by five different values for t, but there are a host of others. For instance, if t = 3.7 hours, 450t would become 450 multiplied by 3.7, which equals 1665 miles. Or, put another way, the plane travels 1665 miles in 3.7 hours of flight. We could do this, in theory, for any of the values of t between 0 and 4. It is as if the expression acts like a seed or computer folder containing all of the information regarding all possible distances of travel from 0 to 4 hours in this scenario.

A second symphony: suppose we want to calculate the amount of money that we would earn when paid an hourly wage of $16 an hour. This generates the following numerical ensemble:

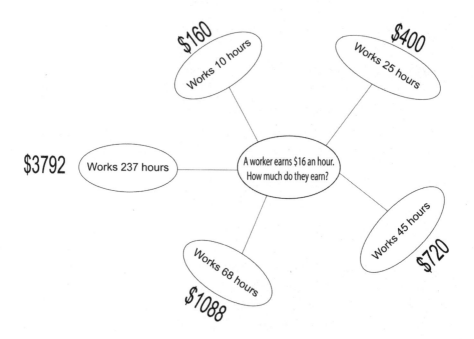

As before, there are a host of other values for wages earned based on the possible hours worked, which could eventually amount to tens of thousands of hours for a given individual. We can store all of these in the following expression, where "h" stands for hours worked:

$$16h,$$

or 16 multiplied by h. For example, all of the earnings in the previous diagram can be easily reproduced by simply setting h equal to each of the following values:

$$h = 10; h = 25; h = 45; h = 68; h = 237.$$

Other everyday situations that yield groups of numbers orbiting a common theme include the following:

- The amount owed on a 30-year house loan of $900,000 after making payments of $5000 for a given number of months. To generate the ensemble, let the number of months vary—then compute how much is still owed after each month.
- The amount of sales tax (at 7%) paid by each individual in a certain town for a given year. To get the ensemble, multiply the total retail spending of each person for the year by 0.07.
- The numerical position of Earth as it travels in its orbit around the sun. To get the ensemble, choose different dates and times throughout the year, then locate the position of Earth.
- The batting average of a baseball player for a given season based on the number of hits obtained in so many at-bats. To get the numerical ensemble, choose the number of hits and number of at-bats for a given player—then calculate $\dfrac{\text{number of hits}}{\text{number of at-bats}}$.

The individual batting average of each of the thousands of MLB baseball players puts a face on many of the values in this numerical symphony (e.g., 0.344 for Miguel Cabrera in 2011, 0.406 for Ted Williams in 1941, and so on).[8]

All of the numerical variation possible in these four scenarios can also be captured and stored with expressions that use letter abbreviations rather than words. The expressions in some cases will be much harder to obtain and more complicated than before, but they still accomplish the same goal.

A NEW WAY OF THINKING

Though numbers have been ever-present in these examples, we are no longer dealing with simple arithmetic. Something more is going on now.

Imagine if you will a child engaged in word play. Such a child might play with words in various combinations and, upon stumbling onto the word *ram* and liking the sound of it, decide to explore further to find other words that sound the same—eventually discovering words like bam, clam, dam, gram, ham, jam, spam, and yam. After this first successful exploration, they may substitute some letters to create new rhymes and eventually learn dozens more new words through this study.

So, the accidental discovery of an unknown word and its pleasant sound has suddenly opened wide, for the child, a whole new way of thinking about words. They can now systematically search for new words that sound alike as well as search for words that have the same meaning.

We are presently at a similar place. But what is it that we have discovered? Is it that we can abbreviate words? Surely there must be more, as abbreviation is not a new technique. In fact, numerals (which go back thousands of years) are themselves shorthand symbols for the quantities they express. And though abbreviation has a significant presence in algebra, it is not what fundamentally differentiates the subject from arithmetic.

What we have inaugurated here that is truly different from arithmetic is a new and deeper way of thinking about certain types of mathematical problems.

For instance, in arithmetic you might be interested in computing your earnings from an hourly wage of $16 an hour. If you worked 40 hours, you would simply multiply 16 by 40 to conclude that you will earn $640. If you work 56 hours another week, then you would calculate 16×56 and move on. Question answered.

What we are doing now is not just looking at a single situation and making a computation, but establishing a rule that holds for the wider variety of situations possible (earnings in this case) and how they relate to one another—think weather on a given date versus climate over a decade. If we can do that, then we can generate any value that we care to know about (e.g., 16h).

The possibilities are immense. Now that we know that some variable phenomena—like money earned from an hourly wage, distance traveled by a plane, and the commutative property of addition—can be readily described by written, abbreviated expressions in this new way of thinking, could it be possible that if we reverse the process and first create abbreviated expressions of our own choosing, then we might eventually be able to describe novel things—presently unknown to us?

For example, we have established that "16h" can be used to help an employee who works h hours at $16 an hour calculate their total wages over a specific period of time. What if we now, from simply looking at this object, decide to find other objects that "rhyme with it" by raising the h to the second, third, or fourth power, obtaining $16h^2$, $16h^3$, or $16h^4$, respectively? Could these new expressions possibly describe some kind of variable behavior that we don't know about yet?

If so, it could be very worthwhile to study these expressions in their own right. This is an exciting prospect as it suggests that we can learn more about sophisticated real-world phenomena by simply studying abbreviated expressions, through the prism of this enlarged outlook, on paper.

However, if we decide to do this, then it leads to an interesting situation. If a car is traveling at the steady speed of 16 miles per hour, we can describe the distance it travels after t hours by 16t. We already know that 16h can be used to describe the earnings of a person who has worked h hours at a rate of $16 per hour. This gives us two separate expressions (16t and 16h) that look different, but we must ask, are they really different?

It turns out that despite their different contexts, they produce the same numerical values when we evaluate them for t = 2, 5, 10 or h = 2, 5, 10. In both cases, we find that we obtain the same numerical values (32, 80, 160) with albeit different interpretations: miles in the first case and dollars in the second.

In fact, this occurs here whenever t and h are set to the same number. So these two expressions in effect generate the same numerical ensemble (in the same way)—meaning that if we divorce them of their interpretations (looking only at the values that they generate from their numerical inputs), they are essentially identical. This scenario will repeat itself with other expressions as well (e.g., "$16h^2 + 70h$" and "$16t^2 + 70t$").

Given that different-looking expressions can produce equivalent values from the same input numbers, you can see how it could be useful and perhaps less confusing, at first, to standardize the letters we use for this purpose. Think of it as putting the letters we use in the same font and size. This will allow us to focus most of our initial attention on how the expressions behave as opposed to being distracted by their appearance.

We do this with language, too. Sometimes we want to be specific and talk about five apples or five cars or five phones, but sometimes we want to be more general (divorcing the objects from any specific interpretation) and simply say we have five things. In business or economics, the term *widget* is sometimes used to represent a generic product.

In mathematics, various names have been given to the unspecified object over the centuries. The medieval Muslims sometimes used the word *shay* to represent unspecified information. Some in India used the abbreviated term *yā*, whereas the Italians of the Renaissance often used the term *cosa*.

Once the idea of systematically abbreviating terms took firm hold in the late 1500s and early 1600s, the unspecified entity took several shapes. One of the earlier suggestions, known as the Viète/Harriot protocol, was that vowels in the Latin alphabet (e.g., A, E, and I) be used; but this idea didn't stick for long.

A later idea employed in the mid-1600s by the French philosopher and mathematician René Descartes was to represent the primary variations in a problem by using letters late in the Latin alphabet ($x, y,$ and z as needed). This is the idea that stuck and is still most often employed in most elementary algebra texts today.

Using this standard means that the expressions we used earlier could translate to the following:

Commutative property of addition	$f + s = s + f$ becomes $x + y = y + x$, where x represents the first number and y represents the second number
Distance traveled by plane	$450t$ becomes $450x$, where x represents the time of travel in hours
Amount earned	$16h$ becomes $16x$, where x represents the number of hours

Notice that it is less distracting mathematically to compare $450x$ to $16x$ than it is to compare $450t$ to $16h$.

So we have in a sense two ways to express and operationalize situations involving quantities that can vary: the generic sense, in which case we generally employ letters such as x, y, and z; and the interpretive sense, where we abbreviate the variable quantities that we want to describe, using whatever letters work naturally. The generic sense is more commonly used when we are doing a general study of how to represent and manipulate variation. By contrast, the interpretive sense is used more commonly when we employ abbreviations to describe a specific scenario.

Let's look at an example of this principle in action. If we want to understand the relationship between the area of any rectangle and its length and width, the standard way to represent the numerical ensemble generated by the interaction of these values would be to use abbreviations for each of the words. We can see this in the formula "area equals length times width," which we shorten to $A = lw$. Though we could write this as $A = xy$, where x and y stand in for length and width, respectively, we rarely do so unless we want to operate on the expression as part of a larger problem where there is some benefit to being more generic.

In most applications (such as physics, engineering, and statistics), x, y, and z are usually avoided and single-letter abbreviations are preferred so that the quantities being related to each other are easier to remember. For instance, in Einstein's famous equation $E = mc^2$, E stands for energy and m for mass, while c follows the universal convention for representing the velocity or speed of light.[9] We will come back to this distinction between standard and interpretive notation in later chapters.

CONCLUSION

We have shown that it is possible to capture, in writing, the essence of many phenomena that vary in value. We can then use abbreviations to further simplify what we have captured with no critical loss of information. This can be thought of as creating a written notation, if you will, for describing on paper numerical phenomena that can change value—just as we already have a written notation for music that allows us to describe on paper something as complicated and varied as the sounds from an hour-long Beethoven symphony.

However, this just barely scratches the surface, for we will soon discover that these written expressions truly distinguish themselves through their dazzling capacity for interacting with each other (and numbers) in ways that allow them to systematically discover unknown facts about the world—like almost nothing else. Taking advantage of this ability for interaction will give us the precise tools we need to completely understand and easily dominate the number of days and age problem.

We shall also find that their capacities for representation, combination, rearrangement, and generalization were ultimately the engines that gave rise to such expressions having an immense expanse all their own—one that has been pivotal in the mathematical, scientific, and technical applications of the last half of the second millennium and on into the third. Called *Hisab al-jabr w'al-muqabala* [calculation by restoration (*al-jabr*) and reduction (*al-muqabala*)] by its ninth-century Persian/Arab father, Al-Khwarizmi, and *The Analytic Art* by its Renaissance European father, François Viète, it is the vast conceptual continent we know today as *algebra*.

2

Art of Maneuver

The mathematics of our day appears to me like a large weapon shop in peace time. The store window is filled with showpieces whose ingenious, artful, and pleasing design enchants the connoisseur. The real origin and purpose of these things, to attack and defeat the enemy, has retreated so far into the background of consciousness as to be forgotten.

—Felix Klein (1849–1925), *Development of Mathematics in the 19th Century*

In walking, it means going around the muddy puddle to get to the store dry and clean. In photography, it entails looking for the best vantage point to make a sunset sing. To sports goers, it can refer to leaving the game early to avoid traffic. To running back Barry Sanders, it meant getting around defenders in ways that were almost choreographic.

Maneuver is all around us in a wide variety of manifestations. In some domains, its presence is so pervasive and overwhelming that its very name simply cannot be hidden from view—taking on the mantle of entire doctrines even. Military thinking is one such arena, where the original meaning of the word was closely allied with the notion of moving forces on the ground into favorable positions that hastened the defeat of the enemy.[1]

In its landmark philosophical document, *Warfighting* (1989), the US Marine Corps gives the following description: "maneuver warfare is a philosophy for generating the greatest decisive effect against the enemy at the least possible cost to ourselves."[2] Methods to achieve this effect now also include deception, surprise, shock, and speed not only on the ground but from the air and from the sea.

Two of the foremost military thinkers in history, Sun Tzu and Carl von Clausewitz, spend attention to the idea in their respective famous works, *The Art of War* and *On War*. It remains a hotly debated and energetic topic in strategic circles.

The word *maneuver*, however, is quite versatile in its other wide-ranging uses and definitions, among them "an action taken to gain a tactical end," "a clever or skillful action or movement," and "doing something in an effort to get an advantage or get out of a difficult situation."[3] Our use of the word in this book will contain aspects of each of these definitions. For our purposes, we summarize with the following general description:

> *Symbolic maneuver includes any introduction, combination, movement, and/or manipulation of symbols (including diagrams) to gain an advantage in knowledge, insight, organization, clarity, efficiency, etc.*

Undoubtedly both too broad a description and simultaneously not comprehensive enough, this definition will serve the good-enough purpose here to characterize what is one of the central and foundational features in algebra. In this chapter, we place a magnifying glass on this cornerstone idea.

MANEUVER IN ARITHMETIC

Symbolic stratagems to gain an edge extend beyond mathematics. The use of symbols in language can also be looked upon as a form of sophisticated maneuver. When we tell someone about a vacation last summer photographing waterfalls in Wells Gray Provincial Park (British Columbia), we are actually using language symbols to get around limitations.

We cannot physically re-create the waterfalls, forests, rivers, mountains, wildlife, adventures, and interactions with people that we experienced on the trip, but we can symbolically share them with others using the words we give them in language—to tell stories of our experiences.

Similarly, many of the techniques and algorithms we employ in elementary arithmetic can be looked upon as symbolic maneuvers.

Consider multiplication: How much in ticket sales might we expect to earn from an event that is scheduled to be attended by 965 people at a cost of $175 per person? We can obtain a reasonable estimate of what the revenue should be by simply finding the answer to 965 × 175. One way to get this would be to take nine hundred sixty-five 175s (one for each attendee) and add them together:

$$(175 + 175 + 175 + \ldots + 175 + 175 + 175)$$

nine hundred sixty-five 175s

Though we can obtain the answer this way, almost no one would do so; it is simply too slow and far too painful. What we generally do is take the information (965 × 175) and maneuver it into another form. A millennium ago, folks might have used some sort of device such as the abacus or counting board to find the answer, but today we have several options.

We could take the numerals and directly key them into a calculator and have the machine do the multiplication for us, using procedures coded in electricity. Or we could reformat the information in a way that allows us to perform swift moves in writing (with the aid of a multiplication table) like so:

$$
\begin{array}{r}
9\ 6\ 5 \\
\times\ 1\ 7\ 5 \\
\end{array}
\qquad \longrightarrow \qquad
\begin{array}{r}
9\ 6\ 5 \\
\times\ 1\ 7\ 5 \\
\hline
4\ 8\ 2\ 5 \\
6\ 7\ 5\ 5\ 0 \\
9\ 6\ 5\ 0\ 0 \\
\hline
1\ 6\ 8\ 8\ 7\ 5 \\
\end{array}
$$

This is a symbolic maneuver that keeps us from having to do anything close to the original 965 additions—conveniently showing us that the revenue should be $168,875.

Symbolic maneuvering also happens when we add fractions with unlike denominators. Consider the addition of one-half to one-third expressed as $\frac{1}{2} + \frac{1}{3}$. Now, we may want to simply add the top two numbers together and the bottom two numbers together to obtain $\frac{2}{5}$, but this value is not accurate in the most common interpretation of fractions.

Think of how much pie you will have if you take half of a pie and add it to a third of another—it is certainly more than two-fifths of the same-sized pie.

A way to obtain the correct value is to transform the two fractions into an equivalent form where they both have the same denominators and to add the top numbers together while leaving the bottom number the same. Doing so in this case will yield denominators of sixths, giving

$$\frac{1}{2}+\frac{1}{3}=\frac{3}{6}+\frac{2}{6}=\frac{3+2}{6}=\frac{5}{6}.$$

This is different from the multiplication example, because converting the problem into a symbolic form is not enough. To keep the ball rolling requires that we further convert the fractions into their equivalents in sixths.

This can be likened to trying to do laundry in a coin-operated machine that only accepts quarters when you have a ten-dollar bill. We have to change the form of the money to make the machine work, but not its value or worth, which in this case translates to 40 quarters.

Let's finally suppose that we're asked to add up all of the counting numbers from 1 to 1000 (1, 2, 3,…, 499, 500, 501,…, 999, 1000) to obtain their total. Straightforwardly, this looks like

$1 + 2 + 3 + 4 + … + 499 + 500 + 501 + … + 997 + 998 + 999 + 1000.$

This would be quite a bit of work even with the assistance of a calculator. However, with a little bit of maneuvering, we can rearrange and rewrite the problem as

$$
\begin{array}{cccccccccc}
1 & + & 2 & + & 3 & + & 4 & + & … & + & 497 & + & 498 & + & 499 & + & 500 \\
+ & 1000 & + & 999 & + & 998 & + & 997 & + & … & + & 504 & + & 503 & + & 502 & + & 501
\end{array}
$$

In this form a symmetry is made bare, which shows us that if we add vertically first, as opposed to horizontally, we obtain repeated copies of 1001 (e.g., 1 + 1000 = 1001, and so on):

$$
\begin{array}{ccccccccccccccccc}
1 & + & 2 & + & 3 & + & 4 & + & \cdots & + & 497 & + & 498 & + & 499 & + & 500 \\
+ \, 1000 & + & 999 & + & 998 & + & 997 & + & \cdots & + & 504 & + & 503 & + & 502 & + & 501 \\
\hline
1001 & + & 1001 & + & 1001 & + & 1001 & + & \cdots & + & 1001 & + & 1001 & + & 1001 & + & 1001
\end{array}
$$

There are a total of 500 copies of 1001 in the addition (you can see this by looking at the entries on the top row from 1 to 500), which means that we can swiftly obtain the answer now by simply multiplying 500 × 1001 to obtain 500,500.

Here, a little maneuvering has turned a problem, which would take longer than 25 minutes for most to directly do even with the help of a calculator, into one that can be done by hand in as quickly as a minute.[4]

These examples point to the fact that maneuvering in mathematics is no small thing and can lead to spectacular savings in the time it takes to find answers to certain types of problems. No less significant is the fact that maneuvering symbols (and thus the ideas they represent) can also lead to sensational gains in insight, clarity, organization, identification, and generalization, too!

The types of maneuvers just discussed depend critically on the specific situation at hand. Imagine a scenario in which we could standardize a much larger class of maneuvers, maneuvers that grant us the ability to systematically solve all kinds of seemingly sophisticated and unrelated problems—enabling us to convert some of the elegance and magic of mathematical ingenuity into routine. In a sense, this is what algebra injects into the mathematical bloodstream: providing a method to reduce the brilliant and extraordinary into the ordinary and reproducible.

Many sixteenth-century mathematicians were simply awestruck by this gift that had fallen into their hands. François Viète was so taken with the sweeping possibilities of algebra that he stated, "The analytic art…claims for itself the greatest problem of all, which is: To solve every problem."[5] Another founding father of modern algebra, Girolamo Cardano, called algebra a "truly celestial gift" and boldly proclaimed that "whoever applies himself to it will believe that there is nothing that he cannot understand."[6]

Famed twentieth-century college basketball coach John Wooden (ten national championships at UCLA) spoke of the phenomenon, of turning the sensational into the routine, in the context of his sport. Not a fan of using emotion and other devices to rise to the occasion in a game, Wooden preferred his teams to achieve a consistently high level of excellence through self-discipline, intelligence, and hard practice. His philosophy was to "let others try to rise to a level we had already attained."[7]

Let's now see how learning the art of maneuver in algebra makes it possible to attain a consistently higher level of excellence in our mathematical game.

APPLES AND ORANGES

If we add 3 apples to 4 oranges, what do we get: 7 apple-oranges? Clearly not, but if we add 3 apples to 4 apples, we do get 7 apples. In each case, we can physically join the collections together to obtain seven distinct objects: apples and oranges in the former case and just all apples in the latter. So why can we simplify the latter description and not the former? What's the difference?

One key difference is that "apple/apples" can refer to any number of apples, and this allows us to absorb two separate descriptions (3 apples and 4 apples), which differ only in number not type of fruit, into the single description of 7 apples. Conversely, apples and oranges are fundamentally different fruits and no such absorption (combining 3 apples and 4 oranges into a single numerical description) is possible if we want to retain their distinction.

So, the fundamental fact that apples and oranges are two different types of fruit is encoded symbolically by the fact that their numerical descriptions can't be absorbed into a single one.

In adding or subtracting the lettered expressions that represent numerical variation, we will be confronted by similar circumstances (from now on we will call such expressions varying, variable, or algebraic expressions). This will turn out to be both a great strength of algebra, in giving the subject its wide scope for handling the mixing of different and same types of behavior, and a great weakness, in that these new rules of operation can be overwhelming in education.

Also, because our native powers of recognition won't be as automatically kind to us as they are in distinguishing different kinds of fruit, it will be harder to initially sift out and work with the various types of objects we will encounter. Let's begin our investigation.

Can we simplify either of the following expressions: (a) $3x^2 + 4x$ or (b) $3x + 4x$?

Before answering, let's first discuss what we mean by the phrase "simplify the expression." To return to apples, we were able to simplify "3 apples + 4 apples" to "7 apples": meaning that two descriptions/terms (3 apples, 4 apples) combine to become a single term (7 apples). In working with different fruits, we were not able to symbolically combine the two terms (3 apples, 4 oranges) into a single term without losing essential information.

In the algebraic cases, by "simplify" we mean can we combine either of the sums involving x's and x^2's into a single term. This is a tall order, for remember that, unlike fruit, both of these expressions can take on a myriad of different values, so the simplification must be equal to the original expression for each and every value that x can represent.

On a first attempt, one might try to combine the expression $3x^2 + 4x$ into $7x^3$: a common student choice. However, consider the following table outlining the results of these expressions for the four values 0, 1, 2, and 3:

Set x to:	$3x^2 + 4x$	$7x^3$	Same Value?
0	$3 \cdot 0^2 + 4 \cdot 0 = 3 \cdot 0 + 0 = 0 + 0 = \underline{0}$	$7 \cdot 0^3 = 7 \cdot 0 \cdot 0 \cdot 0 = 7 \cdot 0 = \underline{0}$	Yes
1	$3 \cdot 1^2 + 4 \cdot 1 = 3 \cdot 1 + 4 = 3 + 4 = \underline{7}$	$7 \cdot 1^3 = 7 \cdot 1 \cdot 1 \cdot 1 = 7 \cdot 1 = \underline{7}$	Yes
2	$3 \cdot 2^2 + 4 \cdot 2 = 3 \cdot 4 + 8 = 12 + 8 = \underline{20}$	$7 \cdot 2^3 = 7 \cdot 2 \cdot 2 \cdot 2 = 7 \cdot 8 = \underline{56}$	No
3	$3 \cdot 3^2 + 4 \cdot 3 = 3 \cdot 9 + 12 = 27 + 12 = \underline{39}$	$7 \cdot 3^3 = 7 \cdot 3 \cdot 3 \cdot 3 = 7 \cdot 27 = \underline{189}$	No

Note that the dot ($\cdot$) is a streamlined symbol for multiplication that avoids the potential confusion of the cross symbol ($\times$) with x. Also, the exponent x^2 means $x \cdot x$ and x^3 means $x \cdot x \cdot x$.

The table shows that the two expressions give the same values when x is 0 or 1, but different values when x is 2 or 3. This means that these two expressions are not equivalent, and thus we can't faithfully preserve all of the information stored in $3x^2 + 4x$ by simplifying it to $7x^3$.

This signifies that, mathematically, $3x^2 + 4x \neq 7x^3$ for most values of x (where "$\neq$" means "is not equal to"). It can be shown in a similar fashion that $3x^2 + 4x$ can't be simplified to other expressions such as $7x^2$, either. The variables x^2 and x are like our apples and oranges—they represent fundamentally different types of variation.

What about the expression $3x + 4x$? Can it simplify to $7x$? Let's construct another table like the previous one and see what happens:

x	$3x + 4x$	$7x$	Same Value?
0	$3 \cdot 0 + 4 \cdot 0 = 0 + 0 = \underline{0}$	$7 \cdot 0 = \underline{0}$	Yes
1	$3 \cdot 1 + 4 \cdot 1 = 3 + 4 = \underline{7}$	$7 \cdot 1 = \underline{7}$	Yes
2	$3 \cdot 2 + 4 \cdot 2 = 6 + 8 = \underline{14}$	$7 \cdot 2 = \underline{14}$	Yes
3	$3 \cdot 3 + 4 \cdot 3 = 9 + 12 = \underline{21}$	$7 \cdot 3 = \underline{21}$	Yes

This time, we see that the two expressions are equal for all four given values of x. In fact, it turns out here that the original expression and the simplified expression are equal for all values of x.

So in this case, the expression $3x + 4x$ and the result $7x$ are interchangeable because the final numerical value obtained from evaluating $3x + 4x$ can be faithfully preserved. We've added apples to apples.

Our task now is to figure out a more efficient way to determine when we can combine terms and when we cannot. What are the criteria? This will be obvious to some of you, but to many it may not be so obvious, and it is worth a bit more discussion: so they too may acquire a firmer grasp of the essential principle.

One way to conceptually sift out the key ingredients is by envisioning the variable terms as more familiar objects. Imagine a set of coins that use variables as their face values, represented as

x by $\left(x\right)$

and

x^2 by $\left(x^2\right)$

Looking at the expressions $3x^2 + 4x$ and $3x + 4x$ in this way gives

$3x^2 \; + \; 4x$ as $3\left(x^2\right) \; + \; 4\left(x\right)$

and

$3x \; + \; 4x$ as $3\left(x\right) \; + \; 4\left(x\right)$

This perspective shows that we have two types of coins: $\left(x\right)$ and $\left(x^2\right)$. As with the fruit, we can simplify an expression involving coins of the same type, whereas with coins of a different type we cannot:

$3\left(x^2\right) \; + \; 4\left(x\right)$ cannot be simplified
to a single term

and

$3\left(x\right) \; + \; 4\left(x\right) \; = \; 7\left(x\right)$

These results match our earlier conclusions using tables. From this, we can immediately surmise that the value of the exponent is an important factor in determining whether we can combine two terms to become one. Although the variables involved both contain the letter x, it appears that the exponent must be the same in both terms or we will have different types of coins.

Let's now consider the case of $6x^3 + 7y^3$. Here, the exponents are the same, but rendering them as coins still shows them to be two independent and different types of objects (when included in the same expression):

$6\left(x^3\right) \; + \; 7\left(y^3\right)$ cannot be simplified
to a single term

So, though it is true that the exponents have to be the same, it seems that the type of variation (represented by letters) needs to be the same, too.

Let's add a little variety to the mix by looking at the following two expressions each containing two terms: (a) $12x^2y^3 + 17x^3y^2$ and (b) $12x^2y^3 + 17x^2y^3$. Converting the variable parts to coins and simplifying where possible yields

$$12\,\widehat{(x^2y^3)} \;+\; 17\,\widehat{(x^3y^2)} \qquad \text{\footnotesize different types of coins cannot be simplified to a single term}$$

and

$$12\,\widehat{(x^2y^3)} \;+\; 17\,\widehat{(x^2y^3)} \;=\; 29\,\widehat{(x^2y^3)}$$

Thus, we can simplify the second expression, but not the first. Based on these results, we are ready to make the following assertion: Two terms represent the same "fundamental type" of variation if they have the same variables each raised to the same powers respectively (and can therefore be simplified).

We see this at play in (b), where x and y are raised to the same respective powers in both terms and thus we are able to combine them to a single term.

Circling the variable works well as a visual guide in cases like these where we are dealing with exponents with whole number values, and where the letters are written in alphabetical order; however, it is just that—a guide.

Now that we understand the rule, we can make the following simplifications:

$$6x + 4y + 12x + 25y \text{ simplifies to } 18x + 29y;$$

$$30x^2y^5 + 40y^4z^8 + 60x^2y^5 - 25y^4z^8 \text{ simplifies to } 90x^2y^5 + 15y^4z^8.$$

These simplifications are symbolic maneuvers that improve readability and clarity. They are the most standard of the fare in elementary algebra, but another essential type of simplification has so far been left out—a real game-changer.

It comes from a property that is truly one of the unsung heroes of elementary arithmetic, one whose tracks are often cleverly masked in elegant algorithms such as long multiplication and long division. However, in algebra there is no more denying this property its place in the sun.

A HERO UNMASKED

It is a fact of arithmetic that

$$\underbrace{3(4+2)}_{\text{left-hand side}} = \underbrace{3\cdot 4 + 3\cdot 2}_{\text{right-hand side}}$$

(where we interpret 3(4 + 2) to mean 3 times the sum 4 + 2).

This is straightforward to verify because on the left-hand side we have 3(4 + 2), which after adding the numbers inside the parentheses becomes 3(6) or 18, and on the right-hand side we have 3 · 4 + 3 · 2, which becomes 12 + 6 or 18. Here is the long form:

- The expression 3(4 + 2) means three copies of 4 + 2 added together or (4 + 2) + (4 + 2) + (4 + 2).
- Dropping the parentheses gives 4 + 2 + 4 + 2 + 4 + 2.
- Rearranging the values gives 4 + 4 + 4 + 2 + 2 + 2 or

$$\underbrace{3\cdot 4}_{4+4+4} + \underbrace{3\cdot 2}_{2+2+2}.$$

We have attached or distributed the 3 to both the 4 and the 2, and you might recall from arithmetic that this property is often called the distributive property of multiplication over addition, or put more simply the distributive property. It also holds if we replace the addition by subtraction, which yields

$$\underbrace{3(4-2)}_{\text{left-hand side}} = \underbrace{3\cdot 4 - 3\cdot 2}_{\text{right-hand side}} \rightarrow \underbrace{3(2)}_{\text{left-hand side}} = \underbrace{12-6}_{\text{right-hand side}}.$$

Both sides give the value of 6.

This property holds for all real numbers, meaning that another ensemble of infinitely many numerical expressions is thrust upon us. This ensemble includes the following expressions:

$$230(18 + 99) = 230 \cdot 18 + 230 \cdot 99,$$

where both sides equal 26,910 [having two hundred thirty (18 + 99)'s means that we individually have two hundred thirty 18s added to two hundred thirty 99s], and

$$12(11 - 7) = 12 \cdot 11 - 12 \cdot 7,$$

where both sides equal 48.

As in Chapter 1, we can capture the essence of the phenomenon and store the varied information algebraically by using a different letter (x, y, and z) to represent each of the three varying numbers that we have in these expressions like so: $x(y + z) = x \cdot y + x \cdot z$. Conceptually when x is a whole number, having x number of $(y + z)$'s means that we individually have x number of y's added to x number of z's. For instance, if we let $x = 98$, $y = 115$, and $z = 345$, this expression becomes $98(115 + 345) = 98 \cdot 115 + 98 \cdot 345$, both sides of which equal 45,080.

You might ask: Why we would ever want to explicitly use this property? It seems quicker to add the two numbers in the parentheses first and then multiply (e.g., $3(4 + 2) \rightarrow 3(6) \rightarrow 18$) instead of distributing first, multiplying each pair, and then adding (e.g., $3(4 + 2) \rightarrow 3 \cdot 4 + 3 \cdot 2 \rightarrow 12 + 6 \rightarrow 18$). Now if we were only adding apples to apples, such as we do in elementary arithmetic, you might be right, and this property could possibly stay hidden in the background. But in algebra sometimes we can't simplify what is in the parentheses first, because we're frequently dealing with different terms (apples to oranges).

For example, how do we simplify $3(x^2 + x) + 7x^2$? We can't combine the terms inside the parentheses first as they are different types, so the usual arithmetic route will fail us here. However, we can still simplify this expression by using the distributive property:

$$3(x^2 + x) + 7x^2 = 3(x^2 + x) + 7x^2 = 3x^2 + 3x + 7x^2$$

After combining the two terms with x^2, this simplifies to

$$10x^2 + 3x.$$

This simplified expression is easier on the eyes than the original expression. Remember that this simplification simultaneously signifies, in a single expression, infinitely many arithmetic simplifications all at once—one for each numerical value that x can represent.

KEEPING TRACK OF SIGNS

Engaging the distributive property also introduces another hidden difficulty from arithmetic. Once negative numbers enter the picture, the ideas get a little more complicated.

The symbol "–" serves a dual role as a minus sign and a negative sign, indicative of both an arithmetic operation and a part of a number (e.g., appearing when we subtract 6 – 4 and as a negative sign in –4 + 6). The language equivalent would perhaps be to use the same symbol both as a punctuation mark and as a letter in the construction of a word.

This is important because we can think of positive and negative signs as interacting with each other as well as with the operations of addition and subtraction. That is, just like two or more numbers can combine to form one number (e.g., 5 + 8 becomes 13), so can two or more signs combine to form a single sign. This is demonstrated in the case of multiplication here:

- Positive times positive becomes positive, or more compactly $(+)(+) = +$.
- Negative times positive becomes negative, or more compactly $(-)(+) = -$.
- Positive times negative becomes negative, or more compactly $(+)(-) = -$.
- Negative times negative becomes positive, or more compactly $(-)(-) = +$.

Some positive sign interactions are often implicit and not always traceable by symbols.

These interactions are on full display in algebra when we utilize the distributive property to simplify expressions like the following:

$$-6(x + 9) + 12(x - 5) + -4(x - 30).$$

Applying the distributive property like so

$$-6\,(x + 9) \; + \; 12\,(x - 5) \; + \; -4\,(x - 30)$$

yields

$$(-6 \cdot x + -6 \cdot 9) + (12 \cdot x - 12 \cdot 5) + (-4 \cdot x - -4 \cdot 30).$$

Applying the rules of signs given above as well as those from addition/subtraction results in

$$-6x + -54 + 12x - 60 + -4x - -120,$$

which, after two "(+)(−) to −" interactions and one "(−)(−) to +" interaction, becomes

$$-6x - 54 + 12x - 60 - 4x + 120.$$

Collecting together and combining like terms (x's to x's and numbers to numbers) gives

$$\underbrace{-6x + 12x - 4x}\,\underbrace{-54 - 60 + 120},$$

which simplifies to

$$\underbrace{6x - 4x}\,\underbrace{-114 + 120}$$

or

$$2x + 6.$$

Correctly simplifying this expression requires that we keep track of the distributive property and the rules for addition, subtraction, and multiplication of positive and negative numbers as well as the algebraic rules for combining variable terms of like types.

We can eliminate working through several of the steps in the previous calculations if we remember all of the interactions possible between the + and – signs, where the – sign simultaneously serves its dual roles. This is a skill worth developing and becomes a huge advantage once mastered, but it can be very confusing and frustrating until such mastery has been achieved.

A REMARKABLE CANCELLATION

Sometimes all of the variations in a problem can align themselves in ways that lead to remarkable cancellations. Let's consider the simplification of the following expression:

$$5(x + 3) + 6(x^2 - 3x) + -2(x - 5) + 15x + 3(-2x^2 - 8),$$

where the values will distribute as indicated:

$$5\,(x + 3) + 6\,(x^2 - 3x) + -2\,(x - 5) + 15x + 3\,(-2x^2 - 8).$$

Multiplying these out and using the rules for positive and negative signs gives

$$5x + 15 + 6x^2 - 18x - 2x + 10 + 15x - 6x^2 - 24.$$

Collecting like terms together (x^2's together, x's together, and numbers together) yields

$$6x^2 - 6x^2 + 5x - 18x - 2x + 15x + 15 + 10 - 24.$$

Simplifying (including combining $5x + 15x$ to $20x$ and $-18x - 2x$ to $-20x$) yields

$$0 + 20x - 20x + 25 - 24,$$

which gives

$$0 + 0 + 25 - 24$$

or

$$1.$$

This simplification may look routine at first, but it reveals a surprising result: All of the infinite variation in this problem no matter the value substituted for x will always combine and arrange in such a way to ultimately yield the number 1. Let's demonstrate this for a few possible values of x.

For $x = 10$, the original expression becomes

$$5(10 + 3) + 6(10^2 - 3 \cdot 10) + -2(10 - 5) + 15 \cdot 10 + 3(-2 \cdot 10^2 - 8).$$

We can simplify inside the parentheses first because we are working exclusively with numbers, and this results in

$$
\begin{aligned}
&5(13) + 6(100 - 30) + -2(5) + 150 + 3(-2 \cdot 100 - 8) \\
&= 65 + 6(70) - 10 + 150 + 3(-200 - 8) \\
&= 65 + 420 - 10 + 150 + 3(-208) \\
&= 635 - 10 - 624 \\
&= 625 - 624 \\
&= 1.
\end{aligned}
$$

For $x = 20$, the expression becomes

$$5(20+3)+6\left(20^2-3\cdot20\right)+-2(20-5)+15\cdot20+3\left(-2\cdot20^2-8\right)$$
$$=5(23)+6(400-60)-2(15)+300+3(-2\cdot400-8)$$
$$=115+6(340)-30+300+3(-800-8)$$
$$=115+2040-30+300+3(-808)$$
$$=2455-30-2424$$
$$=2425-2424$$
$$=1.$$

We can see in each case that we ultimately end up with the same final result of 1. This will happen for the billions and trillions of other values (infinitely many in fact) that we can substitute for x. Each case uses a unique path to get there, but they all wind up simplifying to 1. Try it for a few more values yourself to get a better feel for what is happening.

Consequently, in doing the routine simplification of this single innocent-looking algebraic expression, we are not just doing the simplification for the two paths that the expression takes when $x = 10$ or $x = 20$. Rather, we are effectively doing, in one fell swoop, the simplifications for all of the infinitely many paths that the expression can ever take on!

Put another way, an infinite ensemble of individual arithmetical moves surprisingly all still have enough in common that the result always winds up yielding the value 1; and the basic rules of algebra allow us to easily communicate this grand fact.

This is an example of algebra making fascinating results look very routine.

We are now ready to put algebra to work to make sense of the number of days and age problem discussed in Chapter 1.

SEPARATING OUT NUMERICAL INTERACTIONS

In January of 1897, the great German chemist and discoverer of the element germanium, Clemens Alexander Winkler (in probably a take on the famous Shakespeare quote), stated: "The world of chemical reactions is like a stage, on which scene after scene is ceaselessly played. The actors on it are the elements."[8]

Related statements can be made of quantitative variations, with one possible adaptation being that the world of numerical variations is like a stage, on which scene after scene is ceaselessly played; the storytellers of it are the algebraic expressions we can construct.

We now want to use this stage to orchestrate an algebraic script that provides insight into the number of days and age problem from the previous chapter. As a refresher, you may want to try it again for a few different values:

Pick the number of days you like to eat out in a week (choose from 1, 2, 3, 4, 5, 6, 7). Multiply this number by 4. Then add 17. Multiply that result by 25. Next add the number of calendar years it is past 2013 (e.g., if the year is 2016, then add 3). Now if you haven't had a birthday this year, then add 1587, but if you have had a birthday this year, then add 1588. Finally, subtract the year that you were born from this.

After all is said and done, you should have a very personal three-digit number. Reading it from left to right, the first digit is the number of times you like to eat out in a week and the last two digits are your age.

Someone running through this problem at a normal pace will generate a fluid sequence of numerical calculations on their way to obtaining their very personal number at the end. A second and third person will do the same thing and so on, each producing a number related to them, but different from hundreds of others. It can seem somewhat magical how all of the various instructions, in the haystack of the many possible outcomes, wind up delivering that very special three-digit number for each individual.

If you read the problem carefully, however, you can see that there are numbers that stay the same in everybody's calculations (constants) and values that can potentially differ from person to person (variables). That is, stability and variation are both present. See the two examples given for this problem in the previous chapter that generated the numbers 580 and 237, and compare them with your own.

When the problem is performed in the usual manner of simply picking a number and going through the procedure, we are unsuspectingly mixing together different types of behavior (namely, stability and specific instances of different types of variation): dissolving them collectively into the symbolic cauldron and losing critical information. That is fine if producing a specific value is all that we are interested in. However, it is the algebraic way of thinking that we are trying to apply to the situation now rather than the arithmetic way, which means that we want a more comprehensive and global understanding of what is happening in the problem—particularly why it generates three-digit numbers that are so personal to the many individuals participating in the process, all of which requires that we be more studied in how and what we mix.

In other words, we need to slow down or separate out the numerical interactions in order to isolate the values that are truly constant and unchanging in the problem from the values that are specific instances of a varying quantity. The question is, how do we do this?

Let's begin by first identifying the constant parts of the problem and then separating them from the varying parts for the case of someone who had a birthday during the current year. The processes that are the same for every person engaged in the problem are multiplying by 4, adding 17, multiplying by 25, and adding 1588.

By contrast, the number of days and the year of birth will vary based on the specific participant. There is also a third value that stays the same for each person in a given year, but varies from year to year: the number of years away we are from 2013.

Thus, there are three quantities that can change from person to person and year to year in this problem, meaning that we can think of each of them as a different type of variation (or metaphorically as a different type of fruit). To indicate this, we will tag each variable with a different letter. To remain consistent with standard algebraic notation, let's set x to represent the number of days the participant likes to eat out, y to represent the year of the participant's birth, and z to represent the number of years it is past 2013. The algebraic rules for combination will then automatically ensure that the different types of variation retain their individuality throughout our analysis of the process and won't dissolve with each other or the stable numbers. Let's see what, if anything, we can learn by reframing the problem in this way.

The following table breaks up the problem into various steps and shows the accompanying expression in symbols, assuming that the person has had a birthday:

	Step of the Procedure	Expression at That Step
Step 1	Pick the number of days you like to eat out in a week (choose from 1, 2, 3, 4, 5, 6, 7)	x
Step 2	Multiply this number by 4	$4x$
Step 3	Then add 17	$4x + 17$
Step 4	Multiply that result by 25	$25(4x + 17)$
Step 5	Add the number of calendar years it is past 2013	$25(4x + 17) + z$
Step 6	If you have had a birthday this year, add 1588	$25(4x + 17) + z + 1588$
Step 7	Subtract the year you were born from this	$25(4x + 17) + z + 1588 - y$

All of the content of this problem is now neatly stored away in the final algebraic expression in Step 7. Look for yourself to see if you can find evidence of the four constant processes that stay the same for each person and the three variable processes that can differ from person to person and year to year. This expression visually captures, in algebraic language, the crux of all of the quantities and processes involved. However, in its present form, it is still difficult to see exactly how the procedure generates the personalized three-digit numbers that it does.

The good news is that this algebraic form is now operational and thus far more clearly possesses something that its language form lacks: maneuverability. This was one of the major goals we set out for ourselves, and we have achieved it! Let's now leverage this maneuverability through applying the basic rules of algebra discussed in this chapter to see what is going on in this problem.

Simplifying from the original expression:

$$25(4x + 17) + z + 1588 - y$$

becomes by way of the distributive property

$$25 \cdot 4x + 25 \cdot 17 + z + 1588 - y,$$

and multiplying by 25 yields

$$100x + 425 + z + 1588 - y.$$

Combining the numbers (425 and 1588) gives

$$100x + 2013 + z - y.$$

The last expression gives us the look that we need. To more clearly illustrate this requires only that we rewrite it as $100x + (2013 + z - y)$.

Let's examine how this expression looks for the example of Abu Kamil given in the last chapter. He likes to eat out five days a week, the current date is November 5, 2030, and since his birthday is February 6, 1950, he has already had a birthday this year. This information allows us to set $x = 5$ (days of the week), $z = 17$ (2030 is 17 years after 2013), and $y = 1950$ (year of birth). Plugging these into our algebraic expression yields that

$$100x + (2013 + z - y)$$

becomes

$$100(5) + (2013 + 17 - 1950)$$

or

$$500 + (2030 - 1950)$$

or

$$500 + 80.$$

Let's take a closer look at 500 + 80 before completing the addition. The value 500 comes from multiplying the number of days Abu Kamil likes to eat out, in this case 5, by 100. This gives zeros for the last two digits. The number 80 represents his age and, once added to 500, gives us his magic number: 580. Thus, the first digit (reading from left to right) corresponds to the number of days he likes to eat out in a week (5) and the last two digits (80) correspond to his age.

This situation will always work for anyone less than one hundred years old. If you satisfy that criterion and have had a birthday in a given year, set the variables in the formula for your own circumstances and see how the scene plays out.

Let's now analyze the formula to see why it works for all cases. Because z is the number of years since 2013, the expression $2013 + z$ gives us the current year. For instance, if the current year is 2021, then it is 8 years since 2013 (meaning z is 8), and $2013 + z$ becomes $2013 + 8$ or 2021, which matches.

Because y corresponds to the year of birth of the participant, the expression $\underbrace{2013+z-y}$ with words becomes

$$(\underbrace{2013+z}_{\text{current year}} - \underbrace{y}_{\text{year of birth}}),$$

or more clearly (current year − year of birth).

For someone who has had a birthday in the current year, this gives the age of the person for that year. For someone less than one hundred years old and 10 years of age or older, this will always be a two-digit number.

Finally, the term $100x$ from the formula will generate the numbers 100, 200, 300, 400, 500, 600, or 700, corresponding respectively to the number of days of the week (x) being equal to 1, 2, 3, 4, 5, 6, or 7. Each of these seven numbers has "memory space" for accommodating the addition of two digits in their final two slots.

Because the ages for persons 10 years of age or older but less than 100 years old are two digits long, their value is preserved perfectly in this addition (just as the 500 preserved the 80 in 580); and the final result satisfies the condition that the first digit will equal the number of days the person said they like to eat out in a week and the last two digits will be their age.

This will also be true for those less than 10 years of age as well if we interpret 00, 01, 02, 03,..., 09 as 0, 1, 2, 3,..., 9 years, respectively.

Why won't the problem work for someone who is at least one hundred years old? It fails in such a case because the participant's age will then correspond to a three-digit number, which will force a change in the first digit of the round numbers 100, 200,..., 700.

For instance, let's say that a 102-year-old person likes to eat out six days in a week. Plugging in their values for x, y, and z and simplifying will lead to 600 + (the age of the participant) or 600 + 102. The 1 in 102 will change the 6 in 600 to a 7, thus giving the three-digit number as 702. This implies that the person likes to eat out seven days a week and is two years old, which we know to be false, and so the conclusion of the problem is no longer accurate. This scenario of a change in the first digit will always occur for anyone one hundred years of age or older.

You'll notice that we've focused our analysis on cases where the participant has already had their birthday in the year in question. A slightly different expression is yielded for someone who has not had a birthday yet. The only difference is that 1587 is added instead of 1588. Our expression then becomes $100x + (2012 + z - y)$. Rewriting the part of this expression in parentheses in words yields

$$2012 + z - y\ ),$$
$$\underbrace{2012 + z}_{\text{previous year}} - \underbrace{y}_{\text{year of birth}}\ ),$$

or (previous year – year of birth). This is the age of someone who has yet to have a birthday in the current year. This will still be a two-digit number for someone younger than 100 years of age, from which the rest of the analysis follows verbatim.

CONCLUSION

What are we to make of all of this?

A relatively simple application of a foundational principle of elementary algebra has allowed us to crystallize a process whose explanation on first sight is anything but obvious to most. This illustrates the potential game-changing ability of algebra to be an effective tool in revealing the hidden structures in numerical procedures in ways that enable us to better understand them.

Before moving on, let's analyze just a bit more this idea of mixing and not mixing instances of variation and stability.

Consider the following sets of information:

- **First Set:** $[5 \cdot 2 = 2 \cdot 5]$; $[6 \cdot 3 = 3 \cdot 6]$; $[8 \cdot 5 = 5 \cdot 8]$; $[(-16) \cdot 4 = 4 \cdot (-16)]$; $[25 \cdot 40 = 40 \cdot 25]$.
- **Second Set:** $[5 + 5 = 2 \cdot 5]$; $[9 + 9 = 2 \cdot 9]$; $[20 + 20 = 2 \cdot 20]$; $[-32 + -32 = 2(-32)]$; $[500 + 500 = 2 \cdot 500]$.

If we complete the additions and the multiplications for both cases, we obtain the following information:

- **Third Set:** $[10 = 10]$; $[18 = 18]$; $[40 = 40]$; $[-64 = -64]$; $[1000 = 1000]$.

It is clear that the information in the third set can come from completing the calculations in either of the first two sets. Ask yourself: Which of the three sets of data conveys more information?

The first offers several instances of a fundamental property of multiplication—namely, the order in which we multiply two real numbers does not change the result we obtain. The second set gives us examples of another fundamental fact of real numbers: If you add a number to itself, this is equivalent to multiplying the number by 2. The third set of data gives several instances of the fact that a number is equal to itself.

Though all three sets are true, the first and second show what path we take to get to the result, therefore conveying to us more structural information about real numbers and how they work. In the third set, all of the detail in the previous two sets gets lost by reducing to the simple statement that a number equals itself. This is obvious information that doesn't tell us anything new. Thus, whereas all three sets of data are true, there is something to be gained by keeping the information in a form that shows the bare relationships, or the true nature of the variation, as opposed to only seeing the ultimate result.

The equalities in the original two sets are, of course, representative of infinitely many similar expressions. In both cases, algebra can be used to show what is changing as well as what is not.

In the first set of information, the things that can vary are the two values being multiplied, while the idea that remains constant is the fact that switching the order of the multiplication doesn't change the result. By simply tagging the two variations with the letters x and y, we are able to capture and store it all with a single expression: $x \cdot y = y \cdot x$. From this form, the commutative property of multiplication is transparent, and the entire ensemble of occurrences of this property can be generated.

In the second set, the things that are changing are the values being added to themselves on the left-hand side of the equals sign and being multiplied by 2 on the right-hand side. The items that remain constant are the addition and the number 2. In this case, we simply tag the single varying number by the letter x and everything else is captured and stored by writing $x + x = 2 \cdot x$ or $x + x = 2x$. To get the specific occurrences in the second set, all we need to do is set x equal to 5, 9, 20, −32, and 500.

These are basic examples of how information can be lost in mathematics by simply performing calculations to the end without thinking about how they work. Sometimes if we want to better understand a phenomenon in a more fundamental way and capture its essence, it is best if we preserve the trail of how some of that information was obtained.

This is a critical difference between the way we think in algebra and the way we usually think in arithmetic, and it explains why algebra is taught subsequently to arithmetic. Algebra is more interested in preserving and analyzing the relationships between numbers and operations, rather than only focusing on the calculation of specific values—it is in those relationships that lie the keys to comprehending the inner workings of the processes and configurations that we seek to better understand.

With its ability for tagging and holding different types of variation in suspended animation (amidst a tsunami of numerical data), algebra is tailor-made for this activity. It is the subject's "default setting," if you will.

As originally presented, the result of the number of days and age problem appears to work like magic, but by using a little bit of algebra, the procedure has now been tamed through a relatively straightforward and simple process. One of the requirements is that we preserve the different variations in the problem by tagging them with letters (creating an algebraic expression), followed by simple maneuvers according to straightforward algebraic rules. This allows us to mix together objects of the same type and keep separate the objects of different types. At the end of it all, we obtain a form of the expression that allows us to completely explain what is going on. It is somewhat remarkable that it all works in such a routine way—so routine, in fact, that the subtleties of what is truly happening can be easily missed.

This is but a basic use of algebraic machinery. There are much more powerful ways to demonstrate and program it. In physics, for example, we can perform experiments analyzing a certain type of behavior observing how it changes as we vary properties such as temperature, mass, or velocity and, in the process, generate an ensemble of values interconnected around a central theme. Physical theory can then be used, with algebra as an essential ally, to find an expression that generates the given values or provides effective estimates of them (i.e., mathematically explains the results of the experiments and makes new predictions).

The same may occur in many of the natural sciences and in other important areas, where statistics, operations research, and data science may be called upon to play key roles.

Thus, algebra is indeed "electrical soil." Once something is placed into algebraic form, it enters the mathematical network and acquires the potential to be maneuvered in ways that does remind one a bit of the way that electricity is used to facilitate the transmission of information.

Think of it, a camera phone can capture a complex physical scene as an electrical pattern (via a digital image) and then transmit this image at breathtaking speed to another phone 1000 miles away—in what amounts to a maneuver via technology that elegantly overcomes the difficulties of communicating at great distances. Similarly, algebra can be so used to first capture a complicated numerical variation as a symbolic pattern, then maneuver this pattern into a repackaged, more readable form to great advantage.

This concludes the first movement of the book, where our goals have been to illustrate how to give a mathematical shape to numerical variations, and then learn how to perform basic maneuvers on these symbols to gain additional information and insight. Hopefully the narrative up to this point has made the purpose of these fundamental ideas a little clearer and has nontrivially hinted at their significance.

Now the time has come to transition to the next phase of our journey, where we will come upon what has traditionally been considered the core of elementary algebra, its very heart and soul even: the art and science of solving equations. Historically, it is in this arena where the maneuvering of expressions to acquire new information became something truly spectacular—ultimately reaching such commanding heights that even to this very day they come darn close to obscuring in many people's minds everything else about the entire subject.

Equations and Motions

By this happy invention, modes of investigation
at once difficult and disconnected, and dependent for
success, in each particular case, on the skill and ingenuity
of the inquirer, and often on accident, are reduced to a
simple and uniform process.

—Charles Davies (1798–1876), *Elements of Analytical Geometry
and of the Differential and Integral Calculus*

3

Numerical Forensics*

The truth must be quite plain, if one could just clear away the litter.

—Miss Jane Marple in Agatha Christie's *A Caribbean Mystery*

The letter is susceptible of operations which enables one to transform literal expressions and thus to paraphrase any statement into a number of equivalent forms. It is this power of transformation that lifts algebra above the level of a convenient shorthand.

—Tobias Dantzig (1884–1956), *Number: The Language of Science*

A planet orbiting around the sun is in a state of constant motion. And while the elliptical loop it travels on remains relatively unchanged over short periods of time, its position on that loop is shifting all the time—creating in the process its own numerical symphony. Like the airplane from Chapter 1 that constantly changes its location over a four-hour period, so too does a planet constantly change its position during its orbit.

If it helps, think of a car traveling at the constant speed of 75 miles per hour on Interstate 15 South from Salt Lake City, Utah, to Las Vegas, Nevada. The route along the freeway is determined but the car's location on that route is constantly changing as it travels from north central Utah to southern Nevada.

All of this means that we now in effect have two dramas playing out before us:

---

* The idea for the chapter title came from the article "The Math Myth That Permeates 'The Math Myth'" (Devlin 2016).

1. *The entire path or course:* The ensemble of all possible locations of the planet or the car (namely, the entire orbital path or the entire stretch of I-15 between the two western cities).
2. *The time when a specific location on the path or course is reached:* The time of year when the planet reaches a specific location on its orbit, or the time after leaving Salt Lake City when the car reaches a specific spot on the freeway.

This "rhymes" with the situation in which we now find ourselves. Consider the variable expression $3x + 7$. Like the expressions we've seen before, this one can store and generate an entire ensemble of values. This is the first drama.

The expression acts like an organizing principle for those values—much like an orbit or a highway does for the respective objects that travel along their paths. A few of the possible values in this ensemble are listed here:

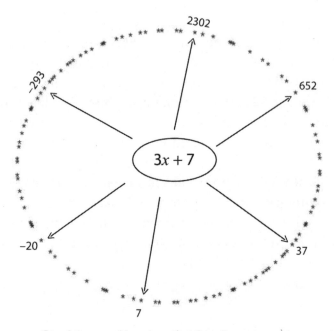

Six of the possible values that $3x + 7$ can generate

Sometimes we are more interested in locating the specific value of the variable x that causes the expression $3x + 7$ to yield a given number. This is the second drama.

Here are the values of x that produce each of the numbers just shown:

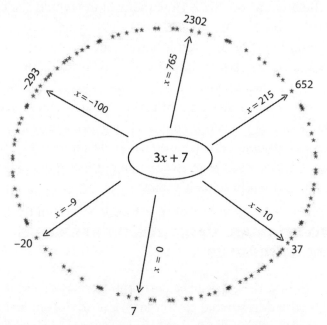

Six specific values of the variable x for the expression $3x + 7$

Metaphorically, the expression $3x + 7$, with its entire ensemble of all possible values, corresponds to the entire orbit of the planet or to the entire stretch of freeway between Salt Lake City and Las Vegas. We can think of all the possible values that $3x + 7$ can take on as the "path" of $3x + 7$, or better yet as the "cloud of possible values" of $3x + 7$.

We have discussed expressions generating ensembles of values in some detail in the first two chapters. This is one side of the coin. Now we want to focus on the other side of the coin, the second drama that involves locating the value of x which causes $3x + 7$ to generate the specific number that we want from the cloud of possible values.

We see this at work in a lot of the problems that we've already encountered. For example, we know that the expression 16h describes the total pay of a person who earns $16 an hour after having worked h hours. In Chapter 1, we chose different values of h to calculate the employee's pay according to how much time they worked, and this allowed us to answer straightforward questions like how much the employee earned if they worked 68 or 237 hours. But what if the employee has a different sort of question? Suppose they have an upcoming expense in the amount of $3792. How many hours would the employee have to work in order to earn $3792?

This type of question is often a harder one to answer and will, in general, require us to perform additional maneuvers: different, in some cases, from those described earlier.

The development and standardization of these maneuvers comprises a huge portion of the "elementary" algebra learned by students up and through calculus. A central goal in this chapter is to understand why the techniques used to answer some of these questions took the forms that are taught in schools today.

THE TWO DRAMAS: VARIABLE EXPRESSIONS VS. BARE VARIABLES

Before we go into the intricacies surrounding the second drama, we need to directly address what can often be a source of confusion. In the case of a planet's orbit or a car on a highway, there is a clear distinction between what the first drama and second drama look like. We list these in the next table.

	First Drama	**Second Drama**
A	Entire elliptical orbit of planet	Time of year when a specific location on the orbit is visited
B	Entire path of I-15 from Salt Lake City to Las Vegas	Time after leaving Salt Lake City when the car visits a certain spot on the freeway somewhere in Utah, Arizona, or Nevada

This distinction becomes less clear when we deal with variable expressions whose ensembles are purely numerical (and not as easy to visualize as an orbit or stretch of road), such as is the case of $3x + 7$ or $16h$.[1]

	First Drama	Second Drama
C	All values stored or generated by $3x + 7$	Specific value of x that makes $3x + 7$ generate a certain value
D1	All values stored or generated by 16h	Specific value of h that makes 16h generate a certain value
D2	All possible hourly earnings from \$16 an hour	Specific number of hours needed to work to yield a particular sum of money

As we move forward, we need to make our analysis of the second drama a bit more operational, so we take note of the fact that the central objects in the first drama are the variable expressions themselves (e.g., "$3x + 7$" or "16h") along with the values that they can produce (metaphorically, the ensembles or clouds of possible values), whereas the focus of the second drama is the search for a specific value of x or h that yields the desired value or location. We can think of the x or the h in these cases as a bare or undressed variable.

Applying this to some of the variable expressions we used earlier yields the following:

Variable Expression	Bare or Undressed Variables
$450x$	x
16h	h
$16h^2$	h
$6x^3 + 7y^3$	x, y
$100x + 2013 + z - y$	x, y, z

Here, the bare variables operationally correspond to the letters (as alphabetic characters) used in each variable expression.

A SIMPLE BUT REVOLUTIONARY IDEA

The central question of this chapter now becomes: Given an algebraic expression (such as 16h, $3x + 7$, or $x^2 - 12x + 100$), how do we find specific values for h or x that cause the expression to yield a given number? The following tables show how we might visualize this question for different desired results:

Algebraic Expression	Question	Value That Works	Verification
$16h$	When does $16h$ yield 64?	$h \to 4$	$16(4) \to 64$
$3x + 7$	When does $3x + 7$ yield 64?	$x \to 19$	$3(19) + 7 \to 57 + 7 \to 64$
$x^2 - 12x + 100$	When does $x^2 - 12x + 100$ yield 64?	$x \to 6$	$6^2 - 12(6) + 100 \to 36 - 72 + 100 \to 64$

A: Expressions, questions, and answers

Algebraic Expression	Question	Value That Works	Verification
$16h$	When does 16h yield 1520?	$h \to 95$	$16(95) \to 1520$
$3x + 7$	When does $3x + 7$ yield 2473?	$x \to 822$	$3(822) + 7 \to 2466 + 7 \to 2473$
$x^2 - 60x + 1200$	When does $x^2 - 60x + 1200$ yield 300?	$x \to 30$	$30^2 - 60(30) + 1200 \to 900 - 1800 + 1200 \to 300$

B: Expressions, questions, and answers

From looking at these expressions, we see that it is a relatively straightforward process to verify that the stated value works once we know what it is—it reduces to a simple calculation. But how do we find the value of the bare variable we're looking for when it hasn't been given to us?

In order to address this question, our first task will be to free it from the shackles of ordinary language and frame it in more maneuverable forms that are hopefully also straightforward to understand. The hope then is that we can learn how to maneuver these forms in systematic ways that allow us to calculate the answer. We saw the power of maneuverability in the previous chapter. It will prove similarly decisive here.

To those who are very comfortable with handling these types of problems, what we are doing may seem like a roundabout way to approach this subject. But history shows us that finding specific numerical values—for bare variables—that satisfied certain conditions was by no means a trivial question. Despite the fact that certain parts of algebra date back to antiquity, it wasn't really until the mid-1500s and into the 1600s, during the rise of the symbolic algebra era, that mathematicians acquired new insights that enabled them to develop more user-friendly methods to answer these questions: methods that allowed both for a comprehensive theory to emerge and for educators to develop systematic ways to teach the subject to the masses.

Let's start by considering the variable expression $x + 4$ with its attendant ensemble of values and asking what value of x will cause the expression to yield the number 14. Arithmetic tells us that this works when x is 10 as $10 + 4$ sums to 14.

Unfortunately, we won't always be able to answer so intuitively questions like these. For instance, can you determine on sight the value of x in $3x + 7$ that will yield 8545? Most likely not. Consequently, we will need to devise techniques that will allow us to project our knowledge beyond the horizon of what we can initially see.

You are already familiar with this type of situation in arithmetic. For instance, though the multiplications $4 \cdot 2$ or $3 \cdot 20$ might be easy mental math for most, this is not the case for products such as $56 \cdot 45$ or $345 \cdot 253$. Techniques learned from our school days, however, allow us to systematically compute, in writing, these latter two products. These techniques require use of a times table and one of the recipes for long multiplication. Once learned, these additional strategies give us the ability to project our knowledge based on the evidence in front of us to produce the information that we want to know and that is presently unknown to us.

This is precisely what we want in the algebraic case as well.

It is easy nowadays to take the equals sign (=) for granted, but throughout most of the history of algebra, it was not used at all. It was introduced to the world in the book *The Whetstone of Witte* (the sharpening of wit), written in 1557 by the Welsh mathematician Robert Recorde. Recorde said he chose the symbol "bicause noe .2. thynges, can be moare equalle" than two parallel lines.[2] The notation didn't immediately catch on—taking nearly a full century more before becoming part of the common mathematical vocabulary.

More important than the sign itself, however (as other signs could have worked just as well), was the revolutionary idea for which it was the "icing on the cake." The idea was that we could place a wide array of problems involving finding or detecting unknown values on a general platform that was much easier to follow than anything else before it—with the crucial benefit that this entire platform could then be operated on in well-scripted and highly visible ways as a mathematical object itself, almost like the very numbers and variables, themselves, that made it up.[3]

It was to prove to be one of the most important operational advancements in the history of all of mathematics, rivaling the introduction of place-value systems into arithmetic—which is saying something.

Using this platform in regard to the question involving finding what value of x makes $x + 4$ become 14, we obtain

$$x + 4 = 14.$$

This object is different from the algebraic expressions in the previous chapter. How so? Well, first is the way in which we choose to interpret it, which is as an algebraic (and more operational) representation of the question involved in the second drama (i.e., for what value of x does $x + 4$ yield 14?). The first drama would be more interested in the general ensemble of values produced by the expression $x + 4$ (not just the value 14) and what the expression can tell us overall about a particular type of variable behavior.

Second, it is an expression that comes as a packaged deal, consisting of two distinct commodities from the start: one commodity "$x + 4$" on the left-hand side of the equals sign and another commodity "14" on the right-hand side. And though these two commodities do not initially seem to interact with one another, we shall soon see that they can be co-opted to work together in concert in ways that will ultimately reveal to us the information we seek.

REPHRASING THE QUESTION

From here on out, we will call expressions such as $x + 4 = 14$ *equations*. The type of equations that we will analyze in this chapter are further distinguished from earlier expressions with equals signs by the fact that they are usually conditional statements—sometimes true and most times false.

To illustrate this, let's look closer at the equation $x + 4 = 14$. It has variation built in it, so in principle we could plug in different values for x. Let's see what it looks like for $x = 0, 3, 10, 25,$ and -20:

Value of x	$x + 4 = 14$ Becomes:	True?
$x = 0$	$0 + 4 = 14$	No
$x = 3$	$3 + 4 = 14$	No
$x = 10$	$10 + 4 = 14$	Yes
$x = 25$	$25 + 4 = 14$	No
$x = -20$	$-20 + 4 = 14$	No

Of the five values tried, 10 is of course the only one that works, and furthermore, of all the infinitely many values that we can plug into this equation, 10 is the only one that makes the equation true—every other one will fail just as 0, 3, 25, and –20 did. There are no other real numbers out there hiding behind a tree that will make both sides of this equation equal.

The fact that there are usually only specific values out of a sea of possibilities that make an equation true means that we'll need to do some investigative work to identify those values.

Values that make an equation into a true statement, as 10 does for $x + 4 = 14$, are called solutions. We would say that 10 is the solution to the equation $x + 4 = 14$, whereas the numbers 0, 3, 25, and –20 are not solutions. Note that 10 would not be a solution to the equation $2x + 8 = 14$ because it produces the statement $2(10) + 8 = 14$, which is false—the solution to this latter equation is the value 3.

Here are a few examples of what rephrasing questions surrounding the second drama into the new platform of equations and solutions looks like:

Variable Expression	Question	Question in Equation Form: Find the Solution to:	Answer in Solution Form
16h	When does 16h yield 64?	16h = 64	h = 4
3x + 7	When does 3x + 7 yield 64?	3x + 7 = 64	x = 19
$x^2 - 12x + 100$	When does $x^2 - 12x + 100$ yield 64?	$x^2 - 12x + 100 = 64$	x = 6

Questions using the platform of equations and solutions

To effectively deal with the second drama, then, requires us to learn how to find solutions to equations. Elementary algebra supplies a most effective recipe for the simplest types.

MOST REMARKABLE PROPERTIES

Let's now take the equation $x + 4 = 14$ and subject it to some arithmetic. We will not play favorites and will abide by the rule that whatever changes we introduce to one side of the equation must be done to the other side as well. The following table shows the results of four such calculations with this equation:

Operation on Original Equation ($x + 4 = 14$) in Words	In Equation Notation	Equation After Simplifying Left- and Right-Hand Sides
Add 20 to both sides	$20 + x + 4 = 14 + 20$	$x + 24 = 34$
Subtract 2 from both sides	$x + 4 - 2 = 14 - 2$	$x + 2 = 12$
Multiply both sides by 5	$5(x + 4) = 14(5)$ or $5x + 5 \cdot 4 = 70$	$5x + 20 = 70$
Divide both sides by 2	$\dfrac{x+4}{2} = \dfrac{14}{2}$ or $\dfrac{x}{2} + \dfrac{4}{2} = 7$	$\dfrac{x}{2} + 2 = 7$

Notice that each operation changes the way the original equation looks—actually converting it into a new equation. Yet, if we replace x by 10 in each of the new equations, we still get a true statement—meaning 10 remains the solution. Remarkably, the solution to the original equation seems completely unaffected by the changes we made!

What would happen if we altered one side of the equation and not the other? If we were to add 20 to the left-hand side of the original equation (obtaining $x + 24$) and leave the right-hand side of the equation as 14, we would create the new equation $x + 24 = 14$. And 10 is not a solution now because $10 + 24 = 14$ is not true.

Try subtracting 2, multiplying by 5, or dividing by 2 only on the left-hand side of the original equation, and you'll see that 10 will fail as a solution in those cases as well. By changing one side of the equation and not the other, we have made an unbalanced adjustment to the equation itself, and this is why it has a different solution.

The properties exhibited in the previous table turn out to be true of equations in general. We state them as a general rule:

> Adding the same number to, subtracting the same number from, multiplying the same number (excluding zero) to, or dividing the same number (excluding zero) out of both sides of an algebraic equation may change the way the original equation looks, but it doesn't change the solution. That is, the solution to the original equation will also be a solution to the new equation.

Think of the solution as the key that opened the original equation and after the changes still unlocks the new equation. This means that there is an entire ensemble of equations that have cosmetic differences in form but turn out not to be different in content—in that they have the same solution. The following diagram summarizes this using our recent results:

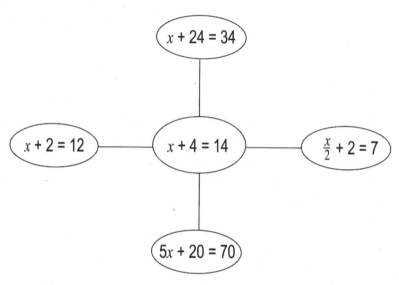

Five cosmetically different equations that have the same solution of 10

There is something truly significant going on here. Remember that each of the equations on the perimeter was obtained from the original equation in the center through a simple arithmetic operation applied to both sides of the equals sign. There are an infinite number of equations that could be generated through arithmetic intervention—for instance, instead of multiplying both sides of this equation by 5, you could multiple both sides by 6, 75, 2000, and so on.

If you perform several of these operations in succession, you can transform a simple equation into one that looks quite different from the original, such as

$$\frac{200x}{7} + \frac{800}{7} + 85 = 485.$$

Yet the solution to this more complicated-looking equation is still 10 because it has been derived from $x + 4 = 14$ using permissible rules (multiply both sides of $x + 4 = 14$ by 200, then divide both sides of that result by 7, and finally add 85 to both sides).

Moreover, if we return to our goal of answering questions about finding specific values for a variable that cause an expression to yield a given result, we see that the following questions relating to the second drama all have the same answer:

- When does the ensemble generated by $x + 4$ yield 14?
- When does the ensemble generated by $x + 24$ yield 34?
- When does the ensemble generated by $x + 2$ yield 12?
- When does the ensemble generated by $5x + 20$ yield 70?
- When does the ensemble generated by $\dfrac{x}{2} + 2$ yield 7?

- When does the ensemble generated by $\dfrac{200x}{7} + \dfrac{800}{7} + 85$ yield 485?

This naturally brings up a question: Because we can convert a given equation into many cosmetically distinct versions of itself, all of which have the same solution, it raises the prospect of whether it is possible to do this in a way such that each conversion gives us a simpler-looking equation than the one before it—culminating, ultimately, to a form in which the solution reveals itself to us for free.

UNDRESSING EQUATIONS

Let's start again with $x + 4 = 14$. But now instead of changing both sides of the equation with random arithmetic, we are going to be more strategic in how we choose to operate, with an eye toward obtaining as the final step the bare or undressed variable. We will for the time being call this method undressing the equation.

In order to get x by itself in this equation, all we need do is get rid of the 4, but we have to do it legally following the rule we've established: What is done to one side must also be done to the other side. Here, simply subtract a 4 from the components on both sides of the equals sign. This yields the following:

Operation on $x + 4 = 14$ in Words	In Equation Notation	Equation After Simplifying
Subtract 4 from both sides	$x + 4 - 4 = 14 - 4$	$x + 0 = 10$, which simplifies to $x = 10$

Indeed, in this case the final step after simplifying results in the value that will make the equation true. The solution won't always be so self-evident, however, so let's now take a look at how this process enables us to find the solution in cases where the value is not immediately clear to us.

Remember our earlier question about finding the value of x in $3x + 7$ that will yield 8545? The equation $3x + 7 = 8545$ is more dressed up, and its solution is not immediately obvious. Undressing it will require two steps this time. We need to free x from both the 7 added to it and the 3 multiplied by it:

Operations on $3x + 7 = 8545$ in Words	In Equation Notation	Equation After Simplifying
1 Subtract 7 from both sides	$3x + 7 - 7 = 8545 - 7$	$3x + 0 = 8538$, which simplifies to $3x = 8538$
2 Divide both sides of simplified equation ($3x = 8538$) by 3	$\dfrac{3x}{3} = \dfrac{8538}{3}$	$1x = 2846$ or $x = 2846$

To further enhance readability, we will now employ the following equivalent vertical format:

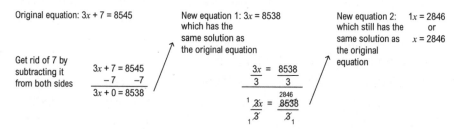

The value of the bare variable gives us the solution to the original equation, and when we substitute 2846 for x in $3x + 7$, we see that it does generate 8545. But something vastly different has occurred now in that we didn't arrive at the solution through trial and error—rather we found a systematic way to carve it out of the original equation! Moreover, we were able to do so in only two maneuvers (getting rid of the 7 from the left-hand side and then getting rid of the 3).

We could have also successfully undressed (solved) the equation by dividing by 3 first and then subtracting 7. However, this would have led to a messier situation involving fractions, so our chosen order of operating is more straightforward.

Another way to look at this procedure of undressing (resolving or solving) an equation is as a process that reduces the original equation to a form in which the solution presents itself as clear as day.

This smells tantalizingly similar in spirit to reducing fractions to simplest form. Consider the following reduction of $\frac{75}{120}$:

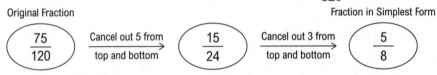

The solving of this equation is a type of symbolic maneuver that has allowed us to project our knowledge beyond what was immediately known to us in order to discover something new. It is different in the details but similar in spirit to our efforts in the first two chapters of simplifying algebraic expressions for greater clarity and understanding.

The equation conveniently contains all of the evidence needed to solve it—that is, all of the information we need is encoded in its structure. All we need do, as Agatha Christie's Miss Marple says, is "clear away the litter" and "the truth will be quite plain." This will generally be the case for the equations that we will solve in this book.

Let's try another equation. Here's an example of how to solve $6x - 2450 = 3694$ using the vertical format:

Original equation: $6x - 2450 = 3694$

Simplified equation 1: $6x = 6144$ which has the same solution as the original equation

Simplified equation 2: $1x = 1024$ which has the same solution as the original equation or $x = 1024$

Get rid of 2450 by adding it to both sides

$$\begin{array}{rcr} 6x - 2450 & = & 3694 \\ +\,2450 & & +\,2450 \\ \hline 6x + \quad 0 & = & 6144 \end{array}$$

Get rid of 6 by dividing both sides by it

$$\frac{6x}{6} = \frac{6144}{6}$$

$$\frac{^{1}\cancel{6}x}{\cancel{6}_{1}} = \frac{\overset{1024}{\cancel{6144}}}{\cancel{6}}$$

The solution is 1024

In practice, much of the captioning is left out without losing computational content, although some conceptual content may be lost. Even the arrows and words are ultimately not needed either, as the right-hand side in the following shows:

Original Equation →	$6x - 2450$	$=$	3694		$6x - 2450$	$=$	3694
	$+\ 2450$		$+\ 2450$	or more compactly →	$+\ 2450$		$+\ 2450$
Simplified eq. 1 →	$\dfrac{6x}{6}$	$=$	$\dfrac{6144}{6}$		$\dfrac{6x}{6}$	$=$	$\dfrac{6144}{6}$
Simplified eq. 2 →	$x = 1024$				$x = 1024$		

Solving these equations then involves choosing carefully crafted arithmetic operations that allow us to isolate the x (or undress it) on one side of the equation. This process of solving the equation by reducing it to its simplest terms was called resolving the equation (or the resolution of equations) in many of the textbooks of the eighteenth and nineteenth centuries.[4]

What happens then, we might ask, if there are variables on both sides of an equation?

VARIABLES ON EACH SIDE

Let's see if we can extend the techniques we've developed so far to solve this equation:

$$15x - 2247 = 12x + 2145.$$

In order to simplify, we will need to maneuver the variable components to the same side of the equation so that we can combine them into a single term. One possible way to go about this is to first subtract $12x$ from both sides, then add 2247 to both sides, and lastly divide both sides by 3. This plays out as follows:

Original equation: $15x - 2447 = 12x + 2145$ | Simplified eq. 1: $3x - 2447 = 2145$

Get rid of 12x
by subtracting
it from both sides

$$15x - 2447 = 12x + 2145$$
$$-12x \qquad\quad -12x$$
$$\overline{3x - 2447 = 0x + 2145}$$

Get rid of 2247
by adding it to
both sides

$$3x - 2247 \;=\; 2145$$
$$+\,2247 \qquad +\,2247$$
$$\overline{3x + 0 \;=\; 4392}$$

Simplified eq. 2: $3x = 4392$

Get rid of 3 by
dividing both
sides by it

$$\frac{3x}{3} = \frac{4392}{3}$$
$$\frac{{}^{1}\cancel{3}x}{{}_{1}\cancel{3}} = \frac{\overset{1464}{\cancel{4392}}}{\cancel{3}_{1}}$$

Simplified eq. 3: $1x = 1464$
or
$x = 1464$

The solution is 1464

Let's verify that 1464 is the solution to this equation by substituting this value for x into both sides of the original equation:

$$15x - 2247 = 12x + 2145$$

becomes

$$15(1464) - 2247 = 12(1464) + 2145$$

or

$$21960 - 2247 = 17568 + 2145,$$

which gives

$$19{,}713 = 19{,}713.$$

Once more, strategic maneuvers of the original equation, this time with an additional step, allow us to systematically solve a problem whose answer is by no means obvious up front. The reduction diagram illustrates this process in an abbreviated fashion:

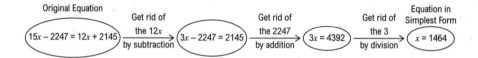

With practice, the mechanical steps to solve equations like these become a matter of routine—so routine, in fact, that it can be easy to forget that the questions we are answering are actually quite sophisticated.

In the case of $15x - 2247 = 12x + 2145$, we have two variable expressions, each generating their own ensemble or cloud of values. For infinitely many choices of x, these two expressions give completely different results:

Value of x	$15x - 2247$ Becomes:	$12x + 2145$ Becomes:	Expressions Equal?
$x = 0$	$0 - 2247 = -2247$	2145	No
$x = 3$	$45 - 2247 = -2202$	$36 + 2145 = 2181$	No
$x = 10$	$150 - 2247 = -2097$	$120 + 2145 = 2265$	No
$x = 500$	$7500 - 2247 = 5253$	$6000 + 2145 = 8145$	No
$x = -20$	$-300 - 2247 = -2547$	$-240 + 2145 = 1905$	No

The question is, for what value of x in the cloud of infinitely many possibilities will these two expressions yield the same result? If we tried this by trial and error, it could take many of us a very, very long time. However, by solving this equation, we were able to obtain the answer to this question in only three steps.

Medieval algebraists could handle questions like this, too. However, they viewed what they were doing in a different way and used words to demonstrate their solutions rather than mathematical symbols. Using modern English words, such a problem might have been posed and solved in writing as follows:

Problem: Given equal amounts of cash in fifteen accounts from which if you remove two thousand two hundred forty-seven dollars you have the same amount as if you had the cash equally distributed in twelve accounts to which you increased that total by two thousand one hundred forty-five dollars, how much cash must there be in each account?

Solution: The computation is thus: Take a given account and call it a thing. From the fifteen take away twelve, which leaves three things. Increase two thousand two hundred forty-seven dollars by two thousand one hundred forty-five, and you obtain four thousand three hundred ninety-two dollars. Now take the four thousand three hundred ninety-two dollars and split it into three portions, which gives one thousand four hundred sixty-four dollars for each portion. This then is the thing or the amount of cash in each account.

This type of presentation of problem solving, where words are fundamental to the demonstration of the solution process, is called *rhetorical algebra*. The substitution of the word "things" for "accounts" here is similar to the process in which we use x to represent the unknown, and it more naturally separates the different types of entities. Dollars can be added to or subtracted from dollars, and things can be added to or subtracted from things.

Our modern platform of equations allows for the separation, in writing, of the solution process from the context in which the problem is delivered. That is, a word problem involving cash in 15 accounts could be written similarly to the rhetorical example above, but our solution method can still employ the x's and numerals we used earlier—making no reference at all to accounts, things, or dollars. It is only at the end, after obtaining the solution, that we can decide to interpret that number in the context of the original problem.

This separation turns out to be both a great strength and a potential weakness of our modern mathematical methods. It is a strength because it allows us to isolate, simplify, and study in great depth the craft of solving equations without overburdening our working memory with the additional weight of interpretations. We can literally turn equation solving into a purely mechanical process with little regard for what each solution means and why we're looking for it.

Yet this very ability to isolate and automate equation solving can become a grave weakness if we do so at the expense of conceptual understanding. In order to foster a greater appreciation of algebra, we need to understand why the equations we're solving matter. And the frustration with the subject many of us feel as students stems from this historical disconnect.

FACES OF THE EQUALS SIGN

In Chapter 1 and Chapter 2, we were more concerned with using the equals sign (=) than understanding its history, but in this chapter we've delved deeper into its introduction in 1557 and the conceptual ideas for which it was, in a sense, the herald. This shift in concern can be looked upon, in a way, as representative of the distinction between the different types of equations and different uses of the equals signs. These different uses can be the source of much confusion for many students of algebra:

	Equation Name	Example	Chapter
1	Conditional Equation	$x + 4 = 14$	Chapter 3
2	a. Identity Equation (with variables)	$3(x^2 + x) + 7x^2 = 3x^2 + 3x + 7x^2$ $= 10x^2 + 3x$	Chapter 2
	b. Identity Equation (with variables)	$x + y = y + x$	Chapter 2
3	Identity Equation (with numbers only)	$8 + 12 = 12 + 8$	Chapter 1

Various types of elementary algebraic equations

CONDITIONAL EQUATION

A *conditional equation* is an equation that is not true for every possible value of x. This type of equation takes its name because its validity is conditional upon a specific value of x that we call the solution to the equation. We've worked with many conditional equations already—as we discovered earlier in this chapter, the only value of x that makes the equation $x + 4 = 14$ true is 10.

As we proceed to solve equations of this type, the goal is to find the numerical value of the bare variable (x = a number). This requires that choreographed arithmetic changes be made simultaneously to both sides of the equals sign until the solution is carved out. How to successfully do this has been the procedural focus of this chapter.

IDENTITY EQUATION (WITH VARIABLES)

Identity equations are equations that are true for every value of x that can be chosen—not conditional based on a specific value of x. You can think of them as representing equivalent expressions or values. For example, the three expressions in $3(x^2 + x) + 7x^2 = 3x^2 + 3x + 7x^2 = 10x^2 + 3x$ are equivalent, in the sense that they are equal for every real number that can be chosen for x. We show this for the examples $x = 0$ and $x = 10$:

$$\underbrace{3\left(x^2 + x\right) + 7x^2}_{A} = \underbrace{3x^2 + 3x + 7x^2}_{B} = \underbrace{10x^2 + 3x}_{C}$$

Set x to:	Becomes:	Expressions Equal?
0	$\underbrace{3\left(0^2 + 0\right) + 7 \cdot 0^2}_{A} = \underbrace{3 \cdot 0^2 + 3 \cdot 0 + 7 \cdot 0^2}_{B} = \underbrace{10 \cdot 0^2 + 3 \cdot 0}_{C}$ or $\underbrace{3(0) + 7 \cdot 0}_{A} = \underbrace{3 \cdot 0 + 0 + 7 \cdot 0}_{B} = \underbrace{10 \cdot 0 + 0}_{C}$ or $\underbrace{0}_{A} = \underbrace{0}_{B} = \underbrace{0}_{C}$	Yes
10	$\underbrace{3\left(10^2 + 10\right) + 7 \cdot 10^2}_{A} = \underbrace{3 \cdot 10^2 + 3 \cdot 10 + 7 \cdot 10^2}_{B} = \underbrace{10 \cdot 10^2 + 3 \cdot 10}_{C}$ or $\underbrace{3(100 + 10) + 7 \cdot 100}_{A} = \underbrace{3 \cdot 100 + 30 + 7 \cdot 100}_{B} = \underbrace{10 \cdot 100 + 30}_{C}$ or $\underbrace{1030}_{A} = \underbrace{1030}_{B} = \underbrace{1030}_{C}$	Yes

We will get the same three equalities of A, B, and C for all other values chosen for x. Try it yourself for $x = 7$ and $x = 25$. You should obtain 511 and 6325, respectively, for A, B, and C.

In the case of identity equations, we are more interested in the relationships between equal expressions than we are in finding specific values of x that make the equation true (as they all already do that). The goal is either to simply state the form of the equivalence ($x + y = y + x$) or to maneuver the entire expression to a simpler form that will aid in clarity and/or usability, as is the case with the simplifications done in the table and with others discussed in Chapter 2.

For the latter goal, we usually start out with a single expression that stands alone without need of an equals sign, then add in equals signs as we begin to simplify it. It is almost as if the initial expression flows into the other simpler expressions, and we use the equals sign to express this flow as well as the equivalence itself.

This is in contrast to conditional equations, where both sides of the equation change together in harmony as we attempt to solve for or isolate the variable.

Sequences of maneuvers that involve simplifying identity equations are sometimes called derivations, especially when they also employ scientific and engineering principles. This is where we can get a chain of many expressions, like the three in the previous table linked together by equals signs. In the sciences and engineering, sometimes well-justified approximations that further simplify expressions may also be involved in the chain of reasoning.

IDENTITY EQUATIONS (WITH NUMBERS ONLY)

For these, a simple arithmetical equivalence or derivation is stated with no variation. These are often single arithmetic statements, such as $1 + 1 = 2$ or $11 \cdot 7 = 77$. Or they can be compound derivations, such as $8 + 7 \cdot 5$, which we often simplify in writing as follows: $8 + 7 \cdot 5 = 8 + 35 = 43$. Another example is $6(9 - 5) + 2(7 + 8)$, which we may write as $6(9 - 5) + 2(7 + 8) = 6(4) + 2(15) = 24 + 30 = 54$. As before, the sense is of one numerical expression flowing into another. We are not making choreographed changes to both sides of the equals signs—our main goal, as with the examples here, is simply to reduce the numerical expression down to a single number.

It is worth pointing out that, with the exception of the fact that an equality of some sort is expressed, there is no universal agreement for using the term *equation* in elementary algebra textbooks. Some definitions include all of these varieties, whereas some only admit the conditional ones.

Since the conditional equations are included in all interpretations, this is the most common interpretation meant when the term *equation* is used.

Often students will see or create an equals sign and presume that they need to solve an identity equation (often called an identity) where no solution is called for. In cases where you aren't sure what to do, try to remind yourself of the context and ask whether this is an equation that is true for all numbers or whether it is true for some numbers and false for others. It may help to test it for a few random cases—just make sure to include choices besides 0 and 1. If it is conditional, then attempting to solve the equation may be appropriate.

Otherwise, all that may be required is to allow one step to flow into the next until a simpler form of the expression is obtained.

The various interpretations of equations and the equals symbol brings to mind polysemous and homonymous words. A polysemous or homonymous word is one that has multiple meanings. The following words have multiple meanings in English:

Bank:[5]
- An establishment for the custody, loan, exchange, or issue of money, etc. (noun).
- The rising ground bordering a lake, river, or sea (noun).
- To bounce (a ball or shot) off a surface (such as a backboard) into or toward a goal (verb).
- To incline laterally (verb).

Table:[6]
- A piece of furniture consisting of a smooth, flat slab fixed on legs (noun).
- A systematic arrangement of data usually in rows and columns for ready reference (noun).
- To decide not to discuss (something) until a later time (verb).

The way we handle such words is by understanding the context in which the word is used, and with practice most users of English have no problems with properly interpreting their meaning. And the same can be said of the various uses of the equals sign. Much as with language, we can learn to interpret that information to understand how the equals sign is being used in a given context.

CONCLUSION

Given the nature of our discussion in this chapter, perhaps it will be useful if we conclude with an example.

Let's say a salesperson contemplating renting a car observes that the costs to rent for 2 days, 3 days, 7 days, and 11 days, respectively, are $70, $90, $170, and $250. If this were you, you might want to know how these prices are generated and their relationship to the rental period. You might also have more specific questions, like what the cost would be to rent a car for a month or how much time you could afford to rent on a certain budget.

The rule can be found in the variable expression $20x + 30$, where x represents the number of days rented.[7] Verify that it gives the correct costs for $x = 2$, $x = 3$, $x = 7$, and $x = 11$.

So, in answering the first question, we now have a variable expression that can generate the cost of any other rental period the salesperson might want to know—that is, it contains the ensemble of all possible rental prices based on the number of days rented. This is the first drama.

All of this assumes, of course, that the rental car company doesn't offer discounts for rentals over a certain number of days. If they do, then another, more complicated expression would be needed to model these changes.

Once we know the formula, we can easily calculate the cost for any rental period. If, for instance, the salesperson plans to rent the car for 28 days, all they would need do is set $x = 28$. This transforms our expression $20x + 30$ into $(20)(28) + 30 = 560 + 30 = 590$, which interprets as $590 to rent the car for 28 days.

If, however, they want to know how much time they can afford within a specific budget, say $900, we'll need to rephrase the question as an equation to identify the specific value of x where $20x + 30 = 900$.[8] This is the second drama. Algebra supplies us with the framework to find the answer by solving the equation $20x + 30 = 900$. The following reduction diagram gives the solution:

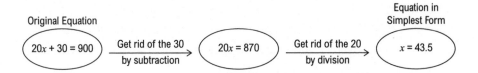

A $900 budget will therefore allow the salesperson to afford a 43.5 day rental, but because half-day rentals are rarely allowed, the longest they can afford to rent the car is 43 days. Note that rounding 43.5 up to 44 days would put the costs $10 over what they can afford.

In many problems that require algebra for clarity, there can be great interplay between the two dramas. Historically, however, it was the second drama of finding specific unknown values that dominated the early history of algebra—with most of it being presented in the rhetorical style using words, as discussed earlier.

It appears that it was only after the advent of more purely symbolic ways to represent algebra that the first drama of systematically representing varying behavior with algebraic expressions was discovered to be an area worthy of study in its own right. This sixteenth- or seventeenth-century breakthrough was critically aided by the spectacular uses of such expressions by scientists, philosophers, and mathematicians, such as Pierre de Fermat, René Descartes, Isaac Newton, and Gottfried Leibniz, who sought to understand and more deeply analyze the problems of motion and the geometrical properties of curves—where they thought of an object's movement as a symphony of changing numerical positions that could be described algebraically or geometrically.

Today, issues relating to the first drama fall under the study of functions—where calculus ultimately came to play a transformational and landmark role. Issues relating to the second drama come under the study of equations—which ultimately led to intellectual advancements every bit as significant and revolutionary as those from the first.

In a very real sense, a great deal of modern mathematics can trace its origins to problems emanating from these two important veins.

With any luck, this chapter has helped you catch a few more glimpses of the vast conceptual continent called algebra. And just as intensely as the subject can mystify and strike dread in the uninitiated, so too can it intensely tempt and mesmerize many a seasoned explorer of its terrain. These polar viewpoints can become highly problematic when the capabilities of algebra and mathematics in general are overestimated, taken on blind faith, or used to intimidate.

4

Converging Streams and Emerging Themes

The essence of knowledge is generalization. That fire can be produced by rubbing wood in a certain way is a knowledge derived by generalization from individual experiences; the statement means that rubbing wood in this way will always produce fire. The art of discovery is therefore the art of correct generalization....The separation of relevant from irrelevant factors is the beginning of knowledge.

—Hans Reichenbach (1891–1953), *The Rise of Scientific Philosophy*

In the background of any area of mathematics lurks a most remarkable and pervasive presence. Mathematician Morris Kline hints at it when he writes:

...A study of mathematics and its contributions to the sciences exposes a deep question. Mathematics is man-made. The concepts, the broad ideas, the logical standards and methods of reasoning...were fashioned by human beings. Yet with this product of his fallible mind man has surveyed spaces too vast for his imagination to encompass; he has predicted and shown how to control radio waves which none of our senses can perceive; and he has discovered particles too small to be seen with the most powerful microscope. Cold symbols and formulas completely at the disposition of man have enabled him to secure a portentous grip on the universe. Some explication of this marvelous power is called for.[1]

Whatever the full extent and nature of this "marvelous power" of mathematics, traces of its presence should be visible even in the most fundamental areas of the subject. As these are the areas where we reside in this book, part of our task in exploring algebra should be to give a glimpse of this potency in its pages. A few observations are the order of the day.

Imagine you knew nothing about mathematics and were playing around with pebbles on the ground. It would still be possible to learn basic facts, such as two pebbles added to four pebbles yields six pebbles, or taking seven groups of three pebbles together yields a total of twenty-one pebbles. In fact, it would be possible to learn a great deal about arithmetic from tinkering around with pebbles only.

But the wonderful thing then is that what is true of pebbles is true of so much more, for if you replace the pebbles with people, dollars, or cars instead, you still obtain true results. That is, it is not just true that 2 pebbles + 4 pebbles = 6 pebbles or that 7 × 3 pebbles = 21 pebbles, but it is also true that 2 people + 4 people = 6 people and that 7 × 3 people = 21 people—and the same with dollars, cars, and so many more things.

In short, by playing with pebbles, one can not only discover arithmetic facts that are true of pebbles, but one can in effect discover arithmetic facts that are true of infinitely many things! It is almost as if the universal laws of addition and multiplication in arithmetic momentarily materialize in the guise of the pebbles in your possession, uncloaking some of their general properties to any and all who would take notice.

Moreover, this remarkable ability to generalize from a specific experience is not unique to mathematics. We may learn the rules for how to safely cross a street on one near where we grew up, yet those same basic principles and behaviors can be used to successfully cross a similar street on the other side of town, in Waterbury (Connecticut), in Trondheim (Norway), or on practically any similar street on the face of the Earth!

From learning how to play an instrument, a board game, a video game, or a sport to learning how to type, drive a car, or cook, this ability to project our knowledge and skills from the context in which we first acquired them, and extend them to millions of other similar circumstances, is everywhere. Amazingly, we all have been granted the ability to connect our individual experiences to something much greater than ourselves—touching, in a sense, eternity itself from the confines of home.

Perhaps the most compelling manifestations of this phenomenon occur in the traditional scientists' and inventors' laboratories, where breakthroughs such as the telephone, the light bulb, X-rays, radioactivity, and the transistor first were discovered or developed in the small, then have ultimately come to affect millions and in some cases billions of people in the large. And this ability to flirt with the eternal from home is also present right here in elementary algebra, where the time has come to spend more of the algebraic capital that we acquired in the first three chapters. Here, we begin with a simple situation from business in the small. Let's see where it takes us.

MONEY STREAMS

Let's consider a situation involving a newly formed small business venture where hamburger meals will be sold. After doing a bit of research, we have determined that overhead (rent, maintenance, utilities, etc.) every two months will cost around $3100. We have also learned that it will cost about $4.75 for the time and ingredients to make each hamburger meal. Based on this information, we have settled on a selling price of $6.00 for each meal. This guarantees that we will make more money from each sale than we spend.

What we would like to know is how many of these meals we have to sell every two months so that the amount of money we bring in (revenue) matches the amount of money we spend (total costs). For the purposes of simplifying the problem, we will make the assumptions that our cost estimates are exact and that we sell all of the food we buy.

To get a feel for the problem, we will start by playing with a few hypothetical situations. What happens, for instance, if we were to make and sell 100 meals? We have two cases to consider: the amount of money brought in and the total amount spent. Based on the conditions just listed, these are both straightforward to obtain:

Revenue (Money Made)	Total Costs (Money Spent)
(Selling price per meal) times (number of meals made and sold)	(Cost to make each meal) times (number of meals made and sold) + overhead
$6 per meal × 100 meals =	($4.75 per meal × 100 meals) + $3100 = $475 + $3100 =
$600	$3575

Situation 1: Make and sell 100 hamburger meals

So, we see in this situation that the revenue is $2975 less than the total costs ($3575 − $600), which is how far we still are in the hole. But this is already an improvement over the $3100 overhead debt we started with before making and selling any meals.

What if we make and sell 400 meals?

Revenue	Total Costs
$6(400) =	$4.75(400) + $3100 = $1900 + $3100 =
$2400	$5000

Situation 2: Make and sell 400 hamburger meals

Here, we see that both revenue and costs have grown, but the gap is closing as the difference between them now is only $2600 versus $2975 from the first situation.

We continue our testing for the situation where we make and sell 1000 meals:

Revenue	Total Costs
$6(1000) =	$4.75(1000) + $3100 = $4750 + $3100 =
$6000	$7850

Situation 3: Make and sell 1000 hamburger meals

Now the gap between revenue and total costs has closed to only $1850. Thus, the two money streams seem to be converging together—with the revenue stream gaining progressively more ground.

We can try to speed things up by using these trials as a springboard to make even better guesses until the desired point is found (where dollars earned equal dollars spent). Many will undoubtedly succeed in this manner. But because we are in the algebraic frame of mind, let's try to use this evidence in a different way to see if we can gain a more comprehensive overview of the forces at work in this problem.

What we have here are two quantities—revenue and total costs—that change depending on the quantity of meals made and sold. These are varying amounts (or numerical symphonies) based on systematic procedures. A few values for each ensemble are given here:

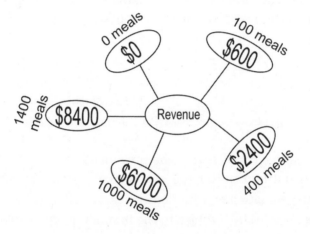

Revenues for number of meals sold

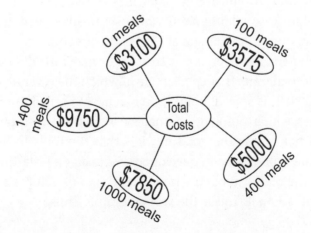

Total costs for number of meals produced (with overhead included)

Can we come up with algebraic expressions that generate each of these numerical ensembles? To do so requires that we separate out the variable components from the constant components. As with the number of days and age problem in Chapter 1, we will do this by tagging the variable terms with letters to allow them to hold on to their separate identities.

Looking at the previous tables and diagrams, we see that the quantities that change from example to example are the number of meals, revenue, and total costs. However, because the variations in the revenue and total costs depend on the number of meals, we will simply tag the number of meals and see what happens. Representing the number of meals made and sold by x gives the following:

Revenue	Total Costs
$6(x) =$	$4.75(x) + \$3100 =$
$6x$	$4.75x + \$3100$

Variable situation: Make and sell x hamburger meals

Notice that we can generate the previous three situation tables by respectively setting x to 100, 400, and 1000 in our variable situation table. We can generate new situation tables as well (e.g., 897 meals sold or 1400 meals sold) by setting x to the number of meals sold. Thus, we have completely captured all of the business situations possible for this hamburger venture—meaning that whatever we decide to do to our variable table, we are doing, simultaneously, to all of the potential sales scenarios in one grand maneuver. This is the first drama.

Our task here, however, is not to store or generate all possible outcomes, but to identify that special one in which the revenue is equal to the total cost. This special point for a business is called the *break-even point*—after which point additional sales in the two-month period produce a net profit. The good news is that the tools to handle this situation have already been developed. We need only elevate this task of ours to the platform of equations and then unleash the techniques from Chapter 3 on it. This is the second drama. Thus,

revenue equal to total costs

becomes in algebraic equation language

$$6x = 4.75x + 3100.$$

Solving this using reduction diagrams yields this:

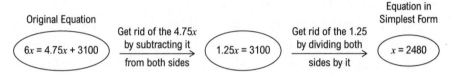

Because the solution to this equation is 2480, we have to sell 2480 hamburger meals in order for revenue to equal total costs over a two-month period. Let's check our work.

Revenue	Total Costs
$6(2480) =	$4.75(2480) + $3100 = $11780 + $3100 =
$14,880	$14,880

Solution situation: Make and sell 2480 hamburger meals

So, we see that once the hamburger meal problem is stripped down to its bare essentials, it lends itself to a very straightforward algebraic treatment. This treatment allows us to almost automatically home in on the answer through the routine solving of a basic equation—swiftly and completely eliminating the need for clever guesswork!

The procedure is so routine, in fact, that it is easy to gloss over what is going on behind the scenes—which is that algebra has allowed us to capture the essence of a task, then gifted us with the ability to efficiently maneuver it in ways, not only saving us from a lot of potential work using trial and error, but also allowing us to build a stage upon which to showcase a strategy for dealing with the entire category of break-even business problems.

A DIFFERENT KIND OF VARIABLE

Just as with the arithmetic rules learned using pebbles, so too does our solution to this break-even problem (learned using hamburger meals) generalize to other scenarios. For as long as the conditions remain the same ($6 selling price, $4.75 cost to make each item, and an overhead of $3100 every two months), it is clearly immaterial if our product changes to chicken meals, packets of writing pens, or any other item. The break-even point will remain 2480 chicken meals, 2480 packets of pens, or 2480 of whatever item we are selling.

This is separating the mathematically relevant components of a problem (selling price and costs) from the mathematically irrelevant details (name and type of product). But this is just the beginning. The algebraic mode of thinking about this problem now comfortably lays before us the potential to bag much larger game. To do so, however, will require the identification of a new type of "variable."

Let's start by going back to the top of our problem to see how revenue and total costs appear in words:

- Revenue = (selling price per meal) times (number of meals made and sold).
- Total costs = (cost to make each meal) times (number of meals made and sold) + overhead.

Since we are now interested in working with the mathematically relevant components, we know (at least as far as the algebra is concerned) that it doesn't really matter what the name is of the product being made and sold. As long as the relationships stay the same, so too does the mathematics. To capture this understanding, we will replace the word "meal" by the more general word "thing." This gives the following:

- Revenue = (selling price per thing) times (number of things made and sold).
- Total costs = (cost to make each thing) times (number of things made and sold) + overhead.

Tagging the number of things made and sold with x gives this:

- Revenue = (selling price per thing) times (x).
- Total costs = (cost to make each thing) times (x) + overhead.

Now for a given business scenario, the number of things made and sold can vary while the selling price, cost per thing, and overhead remain the same (at least over a short period). We saw this happen in our hamburger example—the number of meals varied, while the selling price at $6, cost per meal at $4.75, and overhead at $3100 remained the same. The bigger game we want to bag now is the more general case, where the business scenarios themselves can change (i.e., where the selling price, cost per thing, and overhead change value). We list two such scenarios:

- Selling price per set of knives = $12, cost to make each knife set = $9, overhead = $4000 per month, and x = number of knife sets made and sold.
  - Revenue = $12x$.
  - Total costs = $9x + 4000$.
  - Break-even equation (revenue = total costs): $12x = 9x + 4000$.
- Selling price per purse = $52.25, cost to make each purse = $38.15, overhead = $11,985 per month, x = number of purses made and sold.
  - Revenue = $52.25x$.
  - Total costs = $38.15x + 11985$.
  - Break-even equation (revenue = total costs): $52.25x = 38.15x + 11985$.

Placing the various scenarios in a symphony diagram yields the following:

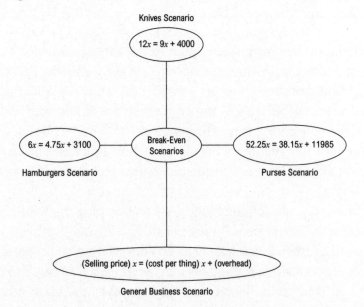

Let's zero in on the "General Business Scenario" at the bottom of the diagram. Notice that, in addition to the x, it has three other quantities that can change: the selling price, the cost to make each thing, and the overhead. These quantities, however, vary in a different way than does the quantity represented by x (the number of things made and sold). These three general quantities stay the same for a specific business scenario such as in the hamburger case, but they can vary from scenario to scenario as in the diagram. Conversely, the x has no such restrictions and can vary within a given scenario, as we saw in the situation tables.

So, the selling price per thing, cost to make each thing, and overhead cost possess both a variable aspect (from scenario to scenario) and a constant aspect (within a given scenario). This makes these three a different sort of variable—a variable that we tune or fix to a certain value for a given scenario, but when the scenario changes, we tune it to other values.

Scenario	Selling Price	Cost Per Thing	Overhead	Break-Even Equation (Selling price)x = (cost per thing)x + overhead
Hamburger	$6	$4.75	$3100	$6x = 4.75x + 3100$
Knives	$12	$9	$4000	$12x = 9x + 4000$
Purses	$52.25	$38.15	$11,985	$52.25x = 38.15x + 11985$

Mathematicians have special names for tuning variables such as these: one of the terms they commonly use is *parameters*. Parameters represent quantitative concepts as well. So, just as we label the regular, less restricted variable (number of meals made and sold) with the letter x, the convention is to represent these parameters by letters, too (often generic ones earlier in the alphabet, but sometimes just a direct, one-letter abbreviation of the quantity name; much more on this later in the book).

For this particular circumstance, we will employ the following abbreviations: selling price = P, cost per thing = C, and overhead = F. In the case of overhead, using an O as an abbreviation could potentially cause confusion with zero, and because overhead costs can also be described as fixed costs like rent and utilities, we have chosen instead to abbreviate the "fixed" with an F for the sake of clarity. If we employ these, our general break-even equation becomes $Px = Cx + F$. And we can rewrite the symphony diagram from earlier with parameters (P, C, and F):

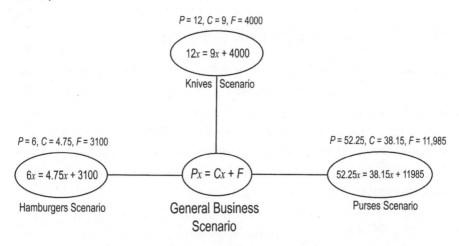

Algebra has lifted us to a high place here. Now when we encounter a break-even problem of this type, instead of using trial and error, we simply ask the following of the problem: What is the selling price per product, what is the cost to make each product, and what is the overhead (or fixed costs)? Then, all we need to do is plug these values into the equation $Px = Cx + F$, solve it, and we are done.

As an example, let's find the break-even point for the purses scenario ($P = 52.25$, $C = 38.15$, and $F = 11,985$):

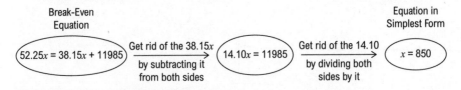

We see for this business model that the point where revenue matches the total costs occurs once 850 purses have been sold. That's it, no situation tables and no guessing. We are done. Period!

And so, algebra allows us to organize and completely dominate this kind of break-even problem. Of course, the details of how to pin down the values for the parameters can become quite involved—especially when inventory, warehousing, and marketing issues are also included—requiring the care and attention of a skilled businessman. But the algebraic way of conducting affairs has allowed us to at least glimpse the soul of the problem and maneuver it to great advantage.

OTHER EXAMPLES OF PARAMETERS

Parameters are one of the most useful devices in the toolkit of anyone trying to capture an entire category or classification of varying behavior. Let's take a look at a few more examples.

Consider three towns with different sales tax rates of 5%, 7%, and 8.2%, respectively. The following diagram shows expressions for calculating the sales tax in a given town, for a given amount of money spent (represented by x):

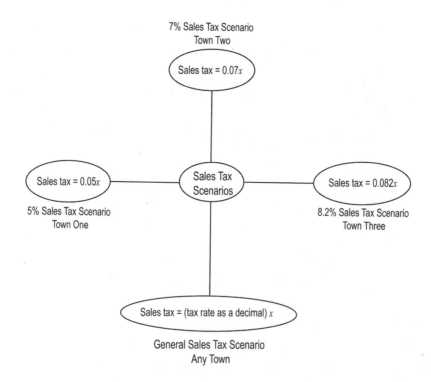

7% Sales Tax Scenario
Town Two

Sales tax = $0.07x$

Sales tax = $0.05x$

Sales Tax Scenarios

Sales tax = $0.082x$

5% Sales Tax Scenario
Town One

8.2% Sales Tax Scenario
Town Three

Sales tax = (tax rate as a decimal) x

General Sales Tax Scenario
Any Town

For example, if you buy $3000 worth of goods in Town Two, then $x = 3000$ and the sales tax you pay is $(0.07)(3000) = \$210$.

Here, we see that the sales tax rate acts as a parameter. It is constant in a given town (over a period of time, at least) but varies from town to town (scenario to scenario). The amount of money we spend can, of course, vary within a given town, and so we think of this as the regular variable and represent it by x.

Though we already have a general formula for calculating the sales tax for any town, we can make it more user-friendly by representing the tax rate by a letter. We make the obvious abbreviation: Tax rate as a decimal = r. This makes our general expression become sales tax = rx. Now that we have conveniently captured the sales tax due on any amount of goods bought in any town, all we need do is substitute the value of the parameter r for a given town and we are ready. In these examples, r would equal 0.05, 0.07, and 0.082, respectively, for the three towns.

We encounter another parameter in physics when we want to calculate the speed of an object dropped from a certain height. After a great deal of experimentation, Galileo theorized that freely falling objects sped up at the same regular rate. In modern parlance, their speed (in feet per second) can be approximated well by the expression $32t$ (32 multiplied by the time in seconds after being dropped). Neglecting the effects of air resistance (also called drag or friction), this means that after 4 seconds of falling, a freely falling object will be traveling at approximately 32(4) = 128 feet per second, and after 7 seconds will be traveling at approximately 32(7) = 224 feet per second, and so on.

Galileo's experiments, though more numerous and far more involved, are in some ways similar to the hypothetical situations that we tested in our hamburger meals break-even scenario. And just as we did there, it is possible to explain the results of his experiments with algebraic expressions.

Moreover, as science progressed into the 1600s, visionaries such as Isaac Newton realized that the algebraic expression for approximating the speed of a falling object on Earth was a special case of a more general situation. Objects on other planets should fall in a similar way as they do on Earth, just at different speeds. Thus, similar but distinct formulas should exist to describe these speeds, too. The following table lists formulas for a few bodies, ignoring the effects of air resistance where necessary:

Celestial Body	Formula for Approximate Speed (of a Dropped Object)	$t = 7$ sec.	Approximate Speed When $t = 7$ Seconds
Earth	32.2t	32.2(7)	225 feet per second (~153 mph)
Venus	29.1t	29.1(7)	204 feet per second (~139 mph)
Mars	12.2t	12.2(7)	85 feet per second (~58 mph)
Moon	5.3t	5.3(7)	37 feet per second (~25 mph)
Jupiter	85.1t	85.1(7)	596 feet per second (~406 mph)

Speed of falling objects on different celestial bodies (these expressions are useful for heights above the surface of body that are small relative to the size of the body; modern-day values rounded to one decimal place have been used; Jupiter, a gaseous planet, doesn't have a solid surface as do the others)

From the table, we see that the generalized expression for approximating the speed after t seconds takes the form

$$\text{speed of fall} = (\text{a number characteristic of each celestial body}) \text{ times } t.$$

This numerical characteristic of each celestial body is a parameter that physicists call the acceleration due to the gravity of that celestial body. This parameter is often abbreviated as g. Using this abbreviation, it follows that the speed of an object's fall is close to gt. Here, the notion of parameters changing values as we go from scenario to scenario now translates to the parameter g changing values as we go from planet to planet (or world to world).

CONCLUSION

The great abilities of algebra to decisively organize, maneuver, and generalize information are on full display here. If you recall, we started out with the relatively pedestrian task of finding out how many hamburger meals we needed to sell so that the money brought in equaled the total amount spent. This we have done with the crucial assistance of basic algebra, and now, because of it, we find ourselves in a commanding position.

From relatively humble beginnings, we now have a general equation ($Px = Cx + F$) that allows us to quickly arrange and solve break-even problems of this type from all over the business landscape—totally eliminating (at this algebraic stage) the need for trial and error. Moreover, along the way we have discovered and made explicit use of an entirely new species of objects called parameters. The systematic introduction and exploration of this class of variables is certainly one of the watershed events in a watershed century (the sixteenth) for algebra.

In the minds of many modern historians of algebra, it was François Viète's understanding and explicit use of parameters that was one of the final decisive steps in propelling algebra into the systematic and broad discipline that it is today.[2] Their introduction helped to literally break the subject wide open and expand the scale of its usefulness—turning small algebra into big algebra.

This is why Viète, more than any of the other great sixteenth-century algebraists before or contemporary with him—including Scipione del Ferro, Niccolò Fontana (Tartaglia), Lodovico Ferrari, Christoff Rudolff, Michael Stifel, Rafael Bombelli, Thomas Harriot, Simon Stevin, or even the talented Girolamo Cardano—is sometimes considered to be the father of symbolic algebra.[3]

For another example of the value of parameters as a tool for studying entire categories of equations and their general properties, see Appendix 1, which includes a brief discussion on the quadratic equation.

To a large extent, before the late 1500s, algebra was seen primarily as a tool for determining unknown values starting from known numerical quantities. This is how the great medieval Islamic mathematician Al-Karaji essentially described the subject in the early eleventh century.[4] Problems were more often than not either phrased in ways that were directly about knowns and unknowns from the start, or in ways that could be translated as such. Once so translated, the techniques of rhetorical algebra (and the slowly emerging symbolic algebra) were then employed to solve them.

Yet Viète saw that algebra still had far more to say about the world. What he saw more clearly than his contemporaries or near predecessors (at least as expressed in print) was that there were all sorts of other situations out there that didn't on the surface look anything at all like the typical problems, yet which were in fact amenable to the same mathematical treatment. In other words, he saw that they were really problems about algebra cloaked in camouflage.

This gave him an expansive view of what algebra really was capable of doing—making him realize, armed with his parameters, that it was an extremely general tool that could aid in the understanding of all kinds of investigations (especially geometrical ones). He gave this enlarged view of the subject the name *The Analytic Art*, and modern mathematics was well on its way to impacting the world as never before.[5]

Compare this to the rise of computers in the twentieth century. In the first half of the century, the primary function of a computer was to perform highly technical numerical calculations for computationally intense endeavors, such as those encountered in making calculations for artillery firing tables or in designing the hydrogen bomb. But as time passed, these machines showed themselves to be capable of much, much more. Like mechanical actors, they became capable of imitating all sorts of machines.

Soon computers were not just lightning-fast calculators, they were becoming typewriters with memory (word processors), spreadsheets, databases, flight simulators, road maps, and social media platforms. Eventually, these machines were able to be miniaturized, and their storage capacities enlarged to such an extent that even the average consumer could afford to purchase one and bring this incredible power and worldwide access into their home.

Computers have changed the way we communicate with one another, creating whole new industries and ways of interacting with the world. All of this has made their impact on human affairs orders of magnitude greater than many of their original practitioners ever imagined.

Algebra has been transformed in a similar fashion and scale, from the early days of finding unknowns in basic numerical and recreational problems to the subject forming critical fuel for calculus (and much of modern mathematics) with the subsequent spectacular applications in physics, chemistry, the sciences in general, engineering, economics, finance, and ultimately to the very computer itself. The uses of algebra continue to expand to this very day.

5

The Rule of Dark Position

Throughout its long history, algebra has used such terms as *roots* (*jidhr* in Arabic, *radix* in Latin), *thing* (*shay* in Arabic, *cosa* in Italian, *coss* in German), *as much as so much* (*yāvattāvat* in Sanskrit), *unknowns*, and *variables* in the service of finding the answers to what we presently call word problems (or story problems).[1] Such problems have consistently been the definitive arena in which to demonstrate to students the applicability of elementary algebra. Their continued use is not without controversy (see Chapter 6).

There are the standard types of word problems that include finding numbers with certain properties, piece length problems, motion in water, in air, or on land, finding unknown angle measures, finding unknown ages, finding amounts invested in various interest-bearing accounts, work rate problems, and mixture problems. Then there are others that involve a twist or two on the standard fare.

As a teacher, I often tell my students not to get too down on themselves when they encounter "that next word problem" down the road that trips them up—and that being stumped, and learning how to handle it, is also important to their mathematical education. Furthermore, I tell them that it takes a good while to achieve the fluency necessary to apply algebra in such situations and that even mathematicians, themselves, are never totally safe from being initially puzzled by some word problem out there on the horizon.

It's not unlike walking incident free for years and still having that one moment on that one day when you stumble suddenly, falling flat and hard (papers flying everywhere) in front of your coworkers. It's just the way it is.

So, what then can we hope to accomplish in a brief chapter like this on word problems? Certainly not the operational mastery of such problems; that is way too large a project to attempt here. We won't even cover every type of word problem mentioned above, much less all of the other types out there. Such a project on its own could easily fill a book or two.

Our goals must be more modest than this, yet at the same time not small. As such, we will keep our focus here primarily conceptual as before—visiting only a few interpretive centers along the way—in our continued attempt to give readers more views of the sweeping possibilities out there in the algebraic countryside.

Let's get started.

MERGING TERMINOLOGIES

WORD PROBLEM 1: If triple a number added to 40 gives 253, what is the number?

This is a classic type of exercise that many newcomers to algebra are likely to encounter. Let's first look at it from the viewpoint of the two dramas introduced in Chapter 3.

TWO DRAMAS VIEWPOINT

This problem has variation in it: triple a number added to 40. And like the student trying to guess at the correct number, we can quickly see this as follows:

- Guess that 5 is the number; then triple 5 added to 40 gives 15 + 40 or 55.

- Guess that 10 is the number; then triple 10 added to 40 gives 30 + 40 or 70.
- Guess that 22 is the number; then triple 22 added to 40 gives 66 + 40 or 106.

We can keep trying to pick numbers and hope that we happen upon the one that works. This is reminiscent of the condition we faced in the last chapter with the situation tables for hamburger meals, and we will deal with it the same way by now pivoting to an algebraic mindset.

Let's represent this variation as an algebraic expression. We need to tag the thing that can vary in the problem by a letter, let's say x, which means that triple a number added to 40 translates to $3x + 40$. This expression captures the ensemble of all possible values. This is the first drama. Using the expression to place the guesses in a diagram yields this:

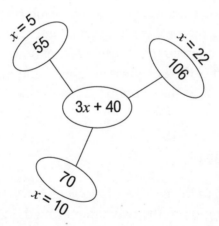

Out of all of the infinitely many values possible, we want the specific number that makes the expression become 253. Sound familiar? This is simply the second drama revisited, which leads to the equation $3x + 40 = 253$. Let's solve this via reduction diagrams:

Original
Equation

Get rid of the 40
by subtraction

Get rid of the 3
by division

Equation in
Simplest Form

$3x + 40 = 253$ $3x = 213$ $x = 71$

Thus, the number that works is 71. Checking our math, $3(71) + 40 = 213 + 40 = 253$.

TRADITIONAL METHOD

If triple a number added to 40 gives 253, what is the number?

Our experience with enough of these problems tells us that this is a problem in which we are trying to find an unknown number, and that the techniques of algebra—namely, setting up and solving an equation—can be employed directly to find this number. So instead of experimenting with a few numbers to get a feel for the situation, we can dive right in and try to set up an equation directly.

The idea is to represent the unknown or the thing you are trying to find by x (other letters will work as well) and then translate the relationship in the word problem into an equation in x. This allows the rephrasing of the question as, "If triple an unknown number added to 40 gives 253, what is the unknown number?" This yields as before the equation $3x + 40 = 253$. The solution follows to yield 71 as the unknown number.

POINTS OF VIEW

Before moving on to the next word problem, it will be prudent to first pay nuanced attention to the two different methods discussed in the previous section. Though they are clearly related, there is a world of difference in how their interpretations play out. The method involving the two dramas may be called the "viewpoint of variable expressions," whereas the traditional method may be called the "viewpoint of the unknown position or value."

To get an appreciation of each method as a distinct viewpoint, let's consider the differences between (1) traveling to an address using local road maps with an eye to better understanding the overall neighborhood and (2) traveling to an address by following instructions from a GPS route finder with the goal of only getting there. The former can give an expansive view of the surrounding region and its place in the greater metropolitan area—allowing the traveler to visualize which roads are connected, see the overall geometry of the road networks, and understand where some of the parks and bodies of water are located—while the latter is strictly concerned with reaching the correct destination.

Both activities may involve some of the same actions—such as traveling on the same street, making the same turns, and waiting at the same stop signs—but the viewpoints and takeaways are totally different.

This approximates some of the differences between the more general "variable expressions" viewpoint and the more specific "unknown values" viewpoint. Both approaches ultimately involve solving the same equation, but the "variable expressions" viewpoint considers the expression $(3x + 40)$ as a generator of a whole terrain of possibilities of which 253 is the one that we are interested in—whereas the "unknown values" viewpoint looks at the word problem as specifying a relationship that we can encode by the equation $3x + 40 = 253$ and then seeks to answer that specific question with less concern about the more general situation. It requires no awareness of any other values that $3x + 40$ can take on.

As we know from Chapters 3 and 4, the "unknown values" viewpoint is closer to the one that dominated much of the early history of algebra—certainly from the time of Al-Khwarizmi in the ninth century up to Viète in the late sixteenth century. This viewpoint has been employed under different guises such as finding "the root," "the thing," or "the cossicke number."[2] In fact, in the earlier to middle parts of the sixteenth century, many for a time called the entire subject of algebra the Cossick Art or the Rule of the Coss, with the practitioners themselves sometimes being called cossists.[3]

Robert Recorde likened the practice of introducing an unknown into the process to the creation of a dark spot in the problem—boldly calling it "the rule of dark position."

Even today, the "unknown values" viewpoint tends to be the predominate method taught in elementary algebra courses. And this is understandable to a degree, since solving word problems by thinking of a letter as storing a piece of unknown information often feels more representative and to the point than does thinking of the letter as storing the possibility of many values—with only one of them being the value we want.

Here, we will subscribe primarily to the "unknown values" point of view, as well. However, it is also useful to keep the "variable expressions" viewpoint in mind, as has been encapsulated in our description of the two dramas. We will freely dial into that more expansive viewpoint when it is convenient and/or illustrative to do so.

DEVELOPING IN THE DARKROOM

WORD PROBLEM 2: A statue is 139 years old. It is four years older than five times the age of an office building downtown. Find the age of the office building.

We'll start by asking, what is it that we are trying to solve for? The problem tells us directly that it is the age of the office building, which is presently unknown to us. We don't, however, let this ignorance stop us and forge ahead anyway by representing the age of the building as x. Injecting this equivalence (age of the building = x) into the word problem makes it become, "A statue is 139 years old. It is four years older than five times x."

Algebra allows us to immediately translate this into "139 years is 4 years older than $5x$" or in symbols alone as $139 = 4 + 5x$ (or $5x + 4 = 139$). Setting up the algebraic equation here is the decisive link, which I sometimes call the "jump for joy stage." Once this stage is reached, the systematic algorithms for solving equations allow us to then maneuver in for the attack. In this case, we subtract 4 from both sides first, which gives $5x = 135$, and then divide both sides by 5, which yields $x = 27$. This tells us that the office building is 27 years old, and we are done.

In this viewpoint of the unknown value, we never looked at $5x + 4$ as a general variable expression producing a lot of different values. We don't care about those other values and didn't consider them. We only cared about the expression as being part of the specific equation $5x + 4 = 139$ and then solving it accordingly.

WORD PROBLEM 3: A 570-foot rope is cut into three pieces. The second piece is four times as long as the first, and the third piece is ten feet longer than twice the first piece. How long are the three pieces?

TRADITIONAL METHOD

Here, we have three unknowns: the lengths of the first piece of rope, the second piece, and the third piece. This may seem like a much harder problem than the ones above it, but we are in luck, owing to the fact that the lengths of the three pieces are related to each other in the following way:

- The second piece is four times as long as the first.
- The third piece is ten feet longer than twice the first.

These relationships mean that we really have to find only one unknown value because once we learn it (the length of the first piece), the other two unknown lengths can be calculated directly from that number. Let's choose to represent the length of the first piece by x. Once we do this, we obtain the following algebraic relationships:

- The length of the first piece = x.
- The length of the second piece = $4x$ (four times as long as the first).
- The length of the third piece = $10 + 2x$ (ten feet longer than twice the first).

Now we just need to find the decisive link—an equation that encapsulates the relationship between the three pieces and the full length of the 570-foot rope. This follows immediately by realizing that adding up the lengths of the three pieces should give us back the 570 feet of rope (assuming no loss of rope in the cutting). Thus, we have

$$\text{length of first piece} + \text{length of second piece} +$$
$$\text{length of third piece} = 570$$

or, using their algebraic names,

$$x + 4x + 10 + 2x = 570.$$

(Jump for joy stage!)

Simplifying the left-hand side of the equation by combining like terms yields $7x + 10 = 570$. Solving this via reduction diagrams gives the following:

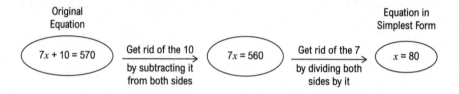

Thus, the first piece of rope is 80 feet long. Using this value sets off a cascade that allows us to find the other two pieces:

- First piece = x, which becomes **80 feet**.
- Second piece = $4x$, which becomes 4 times 80 feet or **320 feet**.
- Third piece = $10 + 2x$, which becomes 10 feet + (2 times 80 feet) = **170 feet**.
- Adding up the lengths of the three pieces gives $80 + 320 + 170 = $ **570 feet**.

We have our answer. Once we set the process in motion, it's as though the three unknowns—the three dark positions—develop right before our eyes. Harvard mathematician Barry Mazur describes his early fascination with this process as follows:

To work out those simple queries…is rather like seeing a concrete visual image develop out of a blank nothing on photographic paper in a darkroom tray. You start with something you deemed X, and at the end of the process you discover X to be concrete, some particular number. There is a sense of power in this (… the early algebraists were very aware of this unexpected power).[4]

TWO DRAMAS VIEWPOINT

Looking at this problem through the lens of the two dramas, we see that the three pieces of rope are related in the following way: the second piece is four times as long as the first and the third piece is ten feet plus twice the length of the first. There is an "ensemble of situations" that satisfy these relationships between a set of three pieces.

For instance, if the first piece of rope is 5 feet long, the second piece would be four times longer (or 20 feet long) and the third piece would be ten feet more than twice 5 feet (or 20 feet long). If we add these three lengths, they combine to form a 45-foot length of rope, which is much shorter than the required length of 570 feet of rope. Thus, it is not the situation that answers the question in Word Problem 3.

Let's represent the stated relationships between the three pieces of rope in the word problem as three bits of information in the following way: (first piece, second piece = four times the first piece, third piece = ten feet plus twice the first piece), or for the specific case in the last paragraph, (5 feet of rope, 20 feet of rope, 20 feet of rope). We will call each set of specific information about the relationships between the three pieces of rope a *triplet*. Because calculating each triplet depends on the length of the first piece, we will denote the first piece as *x*. Doing so yields the following:

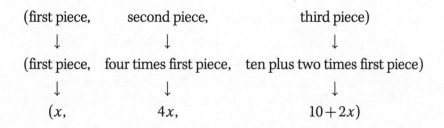

The last one is the variable triplet.

Though this looks different from the single-variable expressions we have seen so far, the variable triplet does the same task, which is to represent (in an algebraic expression) the cloud of all possible triplet combinations. Let's consider a few:

- **First piece is 5 feet ($x = 5$):** $(x, 4x, 10 + 2x) \rightarrow (5, 4(5), 10 + 2(5)) \rightarrow$ (5 feet, 20 feet, 20 feet).

  Sum of the three pieces (length of rope) $= 5 + 20 + 20 = 45$ feet.
- **First piece is 12 feet ($x = 12$):** $(x, 4x, 10 + 2x) \rightarrow (12, 4(12), 10 + 2(12))$ $\rightarrow$ (12 feet, 48 feet, 34 feet).

  Sum of the three pieces (length of rope) $= 12 + 48 + 34 = 94$ feet.
- **First piece is 35 feet ($x = 35$):** $(x, 4x, 10 + 2x) \rightarrow (35, 4(35), 10 + 2(35))$ $\rightarrow$ (35 feet, 140 feet, 80 feet).

  Sum of the three pieces (length of rope) $= 35 + 140 + 80 = 255$ feet.
- **First piece is x feet:** $(x, 4x, 10 + 2x)$.

  Sum of the three variable pieces (variable length of rope) $= x + 4x + 10 + 2x = 7x + 10$ feet.

This ensemble of values summarizes and displays pictorially as follows:

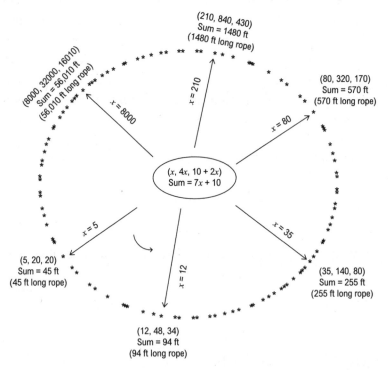

The scenario with $x = 80$ and triplet $(80, 320, 170)$ corresponds to the solution for the conditions described in Word Problem 3.

Now you may be asking: What is the value of looking at the problem this way? It certainly seems a lot longer and more drawn out than using the "unknown values" viewpoint. But what this viewpoint highlights is the capacity of a group of parameters to once again capture an entire class of problems. That is, it shows us that there are infinitely many word problems corresponding to the relationships between the three pieces as stated Word Problem 3. Let's extract a few of them from the ensemble diagram:

- A rope 45 feet long is cut into three pieces. The second piece is four times as long as the first, and the third piece is ten feet longer than twice the first piece. How long are the three pieces? Answer: 5 feet, 20 feet, and 20 feet.
- A rope 255 feet long is cut into three pieces. The second piece is four times as long as the first, and the third piece is ten feet longer than twice the first piece. How long are the three pieces? Answer: 35 feet, 140 feet, and 80 feet.
- A rope 56,010 feet long is cut into three pieces. The second piece is four times as long as the first, and the third piece is ten feet longer than twice the first piece. How long are the three pieces? Answer: 8000 feet, 32000 feet, and 16010 feet.

We've seen similar conditions before. The act of an individual cutting 570 feet of rope into three pieces according to the specifications of the word problem, has infinitely many other situations that "rhyme" with it. It represents a single instance of a symphony of situations that could involve many individuals cutting different-sized ropes into three pieces. Pieces that are related in the same way as described in the word problem. Whether the length of rope is 255 feet or 56,010 feet, algebra allows us to solve them all by setting up the same type of equation. The equations for these two specific situations would respectively wind up as $7x + 10 = 255$ and $7x + 10 = 56010$.

It is worth pointing out that because the length of rope is constant for a particular word problem, but can change value from word problem (scenario) to word problem (scenario), it acts like both a constant and a variable—namely, like a parameter. This means that we can apply big algebra here, if we'd like, and represent the entire class of situations in one grand word problem. Let's represent the length of rope by a letter earlier in the alphabet, say d:

> A rope d feet long is cut into three pieces. The second piece is four times as long as the first, and the third piece is ten feet longer than twice the first piece. How long are the three pieces? The general equation simplifies to $7x + 10 = d$.

We can obtain each of the four word problems in this section from this general one by replacing d by 570, 45, 255, and 56010, respectively.

The "variable expressions" viewpoint has given us a "road map of the neighborhood" for this category of word problems. Moreover, this point of view also hints at the prospect that other word problems we may encounter will have this same circumstance play out—of being a single instance of an infinite assemblage of "rhyming" word problems. It naturally exposes these possibilities. The "unknown values" viewpoint, though solving the problem more clinically, does not as easily reveal the more general outline of these problems.

We will keep the prospect of these more sophisticated numerical symphonies in mind as we move forward.

A TASTE OF GEOMETRY

WORD PROBLEM 4: In a triangle, the measure of the first angle is 5 degrees more than the measure of the third angle. The second angle is 15 degrees more than three times the third angle. Find the measure of each angle.

We are clearly trying to find the measure of the three angles, all of which are currently unknown to us. As the measures of the first two angles are related in a clearly specified way to the measure of the third angle, we will represent this third angle by the unknown x. If we do this, we obtain the following algebraic relationships (reading from the third angle up):

- First angle (in degrees) = $x + 5$ (five degrees more than the measure of the third).

- Second angle (in degrees) = 15 + 3x (fifteen degrees more than three times the third).
- Third angle (in degrees) = x.

The steps to this point have been similar to those in Word Problem 3, but now we're missing something—how to connect this information to an equation. It is implicitly embedded in the problem, but a crucial fact about triangles is needed to see it. What is it?

It is the property that if we add up the degree measures of each of the three angles in any triangle, regardless of shape or size, we will always obtain the same value of 180 degrees.[5] This is a well-known and basic geometric fact about triangles, but it is actually quite remarkable when you consider that there are infinitely many possible triangles out there.

Similar universal truths are known about other familiar shapes—you may know that the circumference of a circle of any size can be approximated by the simple calculation of 3.14 times the diameter, or more exactly by the expression pi times the diameter or πd.

We can use this general property of triangles to set up the equation: first angle + second angle + third angle = 180. Translating into algebraic language gives $x + 5 + 15 + 3x + x = 180$, which when simplified gives $5x + 20 = 180$. The following reduction diagram takes us to the solution:

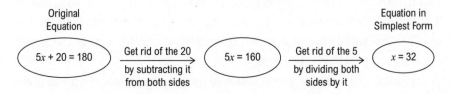

Once we know that the third angle has a measure of 32 degrees, we can calculate the other two (again from the third angle up):

- First angle = x + 5, which becomes 32 + 5 = **37 degrees.**
- Second angle = 15 + 3x, which becomes 15 + 3(32) = **111 degrees.**
- Third angle = x, which becomes **32 degrees.**
- Adding up the three angle measures gives 37 + 111 + 32 = **180 degrees.**

WORD PROBLEM 5: A rectangle has a perimeter of 2434 feet. What are the dimensions of the rectangle if its length is 151 feet less than three times its width?

The dimensions of the rectangle here mean its length and its width. The problem additionally tells us how these two are related with the length being defined in terms of the width. Thus, we will let the width be represented by x. This yields the following:

- Width = x.
- Length = $3x - 151$ (151 feet less than three times its width).

Note that composing our expression for the length requires a little more attention to detail here, because the order in which we subtract values makes a difference in the result. After all, $20 - 8 = 12$ is not equal to $8 - 20 = -12$.

This difference in the order of subtraction is usually simpler to catch in arithmetic due to the fact that the two numbers can be combined right away, but in algebra where different terms in expressions can't always be combined right away, it becomes easier to miss an initial error. That is, if you use $151 - 3x$ for the length instead of the correct form $3x - 151$, there is no immediate alarm bell going off that will allow you to catch the error quickly as there might be with simple arithmetic. The error only becomes apparent at the end of the procedure when you obtain a negative value for x (or the width), which for a physical length should be a positive number.[6]

As before, we need some additional knowledge of geometry in order to create our equation. This knowledge once more is implicitly given in the problem and requires that we know how to calculate the perimeter of a rectangle.

Just as a reminder, the perimeter (circumference) of an object, such as a rectangle (circle), is the distance around the boundary of that object. For the rectangle, we have the following:

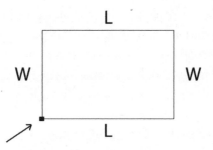

If we start at the point in the lower left corner and walk around the boundary of the rectangle in a counterclockwise direction back to the starting point, we will cover a distance given by L + W + L + W, which simplifies according to the algebraic rules as 2L + 2W. Thus, we have that the perimeter = 2L + 2W.

In the word problem, we know that the perimeter is 2434 feet, and thus we obtain the equation 2L + 2W = 2434. In "x" language this translates through W = x and L = $3x$ – 151 to

$$2(3x - 151) + 2x = 2434,$$

which after multiplying by 2 via the distributive property simplifies to

$$6x - 302 + 2x = 2434$$

or

$$8x - 302 = 2434.$$

Solving this using the standard techniques we've learned gives x = 342. This sets off the cascade:

- Width = x, which becomes **342 feet**.
- Length = $3x$ – 151, which becomes 3(342) – 151 = 1026 – 151 = **875 feet**.
- Checking the perimeter gives 2(875) + 2(342) = 1750 + 684 = **2434 feet**.

These geometrical problems show that sometimes finding a solution requires knowledge outside of what's explicitly stated in the problem. This may seem obvious to say, but it is still useful to keep in mind when tackling these types of situations—as keeping such an awareness on the table can often give one a more flexible approach to solving a problem.

We've also witnessed here a fruitful partnership between algebra and geometry. In the seventeenth century, this partnership would turn into an exciting marriage of the subjects to form the gem known as analytic geometry. And though beyond the scope of this book, this creation of Pierre Fermat, René Descartes, and others would turn out to be a conceptual advancement of the absolute first rank. Algebra and geometry have worked together happily ever since.

PACKAGED CURRENCIES

WORD PROBLEM 6: Given a collection of dimes and quarters worth $34.75, how many of each type of coin must there be if the number of dimes is twelve more than triple the number of quarters?

Here, we are clearly trying to find out how many dimes and quarters we must have to total exactly $34.75, while at the same time having the number of dimes be 12 more than triple the number of quarters.

Though it is certainly possible to have various combinations of quarters and dimes that total $34.75 (137 quarters and 5 dimes, or 111 quarters and 70 dimes, for instance), the task is also to make the combination satisfy the specified relation. The relation in the problem sets as a condition that the number of dimes exceed the number of quarters by a specified amount, and neither of these scenarios satisfy this.

Given that we have many different combinations—of dimes and quarters—to choose from, we will let the "electrical soil" of algebra do its job by helping us ferret out the correct one. However, this problem is different in some ways from the previous problems in this chapter. Dimes and quarters each have a distinct monetary value, so simply knowing the number of dimes and quarters is not enough to know whether they will collectively be worth \$34.75. We have to multiply each coin by its worth (10 cents for each dime and 25 cents for each quarter) to obtain the total amount of money they are worth, so the dimes and quarters here each represent packaged currency.

To get more of a flavor for the problem before we move into how algebra will help us solve it, let's guess that there are 5 quarters. To satisfy the condition of the problem, the number of dimes must be twelve more than triple the number of quarters—meaning we must triple 5 to obtain 15, and then add 12, which gives us a total of 27 dimes.

If this problem were like the rope length or angle problems, then we could directly add 27 + 5 to obtain 32 of something. But 32 of what? Upon closer inspection, this clearly doesn't make sense because we are talking about different types of coins, each with its own currency value. So, before we can test this answer to see if the conditions of the problem are met, we must convert the coins into their monetary values.

In this case, 5 quarters becomes 5 times 25 cents, and 27 dimes becomes 27 times 10 cents, which total 125 + 270 = 395 cents, or \$3.95. Clearly, our guess is off the mark, as this is much lower than \$34.75.

Let's take another crack at it by guessing 30 quarters this time. Here, the number of dimes (twelve more than triple the number of quarters) must be "triple 30 plus 12," or 3(30) + 12 = 102 dimes. The monetary value of these coins is 30 quarters + 102 dimes = 30(25 cents) + 102(10 cents) = 750 + 1020 = 1770 cents or \$17.70, which is a significant improvement on our first guess but still low of the mark.

This time let's use algebraic thinking and guess that there are x quarters. This means that the number of dimes must be "triple x plus twelve" or $3x + 12$. Proceeding as before, the monetary value is now

$$x \text{ quarters} + (3x+12) \text{ dimes} = x(25 \text{ cents}) + (3x+12)(10 \text{ cents}),$$

which after putting the currency values in front becomes

$$25x + 10(3x + 12).$$

This variable expression is equipped to give the monetary values (in cents) of all combinations that satisfy the specified relationship between quarters and dimes in the word problem. Here is the symphony diagram corresponding to our two previous guesses:

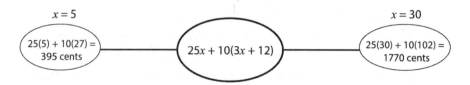

The question of the problem can now be rephrased as follows: Out of all the possible monetary values generated by the expression $25x + 10(3x + 12)$, what is the one that yields a value of 3475 cents ($34.75)? This leads directly to the equation $25x + 10(3x + 12) = 3475$. (Jump for joy stage!) And once we have our equation, the problem is then on the algebraic grid.

The following reduction diagram shows how the rules allow us to maneuver to the solution:

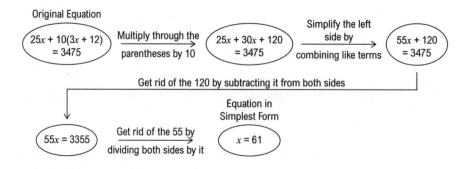

There are 61 quarters, which cascades (via $3x + 12$) to $3(61) + 12$ or $183 + 12 = 195$ dimes.

If we check our work, 61 quarters and 195 dimes give the correct amount: $61(25) + 195(10) = 3475$ cents, or $34.75.

Let's do one more example, this time with paper currency.

WORD PROBLEM 7: A total of $17,252 is divided up into one-, five-, and ten-dollar bills. If the number of fives is seven times the number of ones and the number of tens is quadruple the number of ones, how many of each type of bill are there?

We are trying to find the number of one-dollar bills, five-dollar bills, and ten-dollar bills satisfying the specific relationships given to us by the problem and totaling $17,252. Because the number of fives and of tens are defined in terms of the number of ones, we will let the latter be x. This gives the following:

- Number of one-dollar bills = x.
- Number of five-dollar bills = $7x$ (seven times the number of one-dollar bills).
- Number of ten-dollar bills = $4x$ (quadruple the number of one-dollar bills).

Based on the previous problem, we know that each of these bills is packaged currency and must therefore be converted into their respective dollar values of $1, $5, and $10 when we make the decisive step of placing them into an equation that reflects the situation. Doing so yields

1(number of one-dollar bills) + 5(number of five-dollar bills) + 10(number of ten-dollar bills) = $17252,

which translates algebraically to

$$1(x) + 5(7x) + 10(4x) = 17252.$$

After multiplying, this simplifies to

$$x + 35x + 40x = 17252$$

or

$$76x = 17252.$$

Dividing both sides of the equation by 76 gives us a solution of $x = 227$. We cascade this result to discover the number of five- and ten-dollar bills:

- Number of one-dollar bills = x, which becomes **227**.
- Number of five-dollar bills = $7x$, which becomes $7(227) = $ **1589**.
- Number of ten-dollar bills = $4x$, which becomes $4(227) = $ **908**.

The total dollar value is $1(227) + 5(1589) + 10(908) = \$227 + \$7945 + \$9080 = \$17252$.

CONCLUSION

In the first line of his landmark ninth-century book—*Al-Kitāb al-Mukhtaṣar fī Hisāb al-Jabr wa'l-Muqābala* (*The Compendious Book on Calculation by Completion and Balancing*)—Al-Khwarizmi stated, "When I considered what people generally want in calculating, I found it always is a number."[7] This seemingly simple statement, from the standpoint of today, implied far more diversity in Al-Khwarizmi's era.

Modern views of algebra fixate on the relationships between numbers (constants) and different types of fundamental variations (x, x^3, x^5, and x^8, etc.), but at the time Al-Khwarizmi wrote these words, and all the way up through the sixteenth century, mathematicians thought of fundamental variations as numbers in their own right—just of a different species or type.[8]

Al-Khwarizmi states a few paragraphs later that, for the problems he treats, there are three types (species) of numbers: squares, roots, and simple numbers. He also mentions that these may be joined together to create certain compound species.[9] In the language we have been using, this translates to the three fundamental types: square variations, linear variations, and constants, respectively, represented symbolically by x^2 forms (such as $5x^2$, $-8x^2$, etc.), x forms (such as $76x$, $7x$, etc.), and constants (such as 3, 10, etc.)—complete with their possible combinations ($5x^2 + 7x + 5$, $7x + 10$, $11x^2 + 2x$, etc.).

Interestingly, all of the word problems we have considered in this chapter reduce to expressions involving only x variations and constants. No square variations occur at all. More involved word problems could, of course, require square variations and/or even more complicated ones.

Now in closing, how do we summarize the many strands running through this chapter?

Clearly, the most central of the themes and the one that ties all of the different strands together is the notion of "method." Of method, Alfred North Whitehead states, "The science has not been perfected, until it consists in essence of the exhibition of great allied methods by which information, on any desired topic which falls within its scope, can easily be obtained."[10]

Here, we have problems considering a variety of topics such as rope lengths, angle measures, ages of buildings, number of coin or bill denominations, and the dimensions of rectangles. Yet their resolutions in algebra all reduce to equations of the same basic types involving terms that are multiples of x plus additional constants. Furthermore, standard methods are involved in handling them no matter the original context of the problem. This is information that we might have completely missed out on but for the systematic techniques of algebra.

In other words, even if we had been able to find the answers to each of these word problems through arithmetical ingenuity alone, the techniques so developed for each would have most likely seemed disconnected and haphazard—thus making it hard to generalize them and educate others in their use. However, the methods of algebra unite the word problems in such a way and to such a degree that their commonalities become more transparent, which can help us to gain new insights. Such unification also opens up new horizons for making the subject accessible to larger audiences of people—thus extending the reach of the subject.

To get a feel for what we mean by "method," let's go back to arithmetic for a moment and consider these three problems:

1. Given a collection of 912 objects, how many times can we take away 16 objects from the collection until it is exhausted?
2. Given 912 people, if we organize them into teams of 16 players, how many teams will there be?
3. Given $912, if we partition or distribute this amount equally between 16 people, how much money will each person receive?

These each describe a different action and their answers are interpreted in different ways, but they are all solvable by one and the same method, namely division. To answer each, all we have to do is calculate $\frac{912}{16}$. This can be computed through the process of long division as follows:

$$
\begin{array}{r}
5\ 7 \\
16\overline{\smash{)}\ 9\ 1\ 2\ } \\
-8\ 0 \\
\hline
1\ 1\ 2 \\
-1\ 1\ 2 \\
\hline
0
\end{array}
$$

We obtain 57, which is interpreted, respectively, as 57 times, 57 teams, and $57.

Here, the meaning in each of the three individual problems, briefly, gets turned over to the syntax and maneuvers of long division. And in the same way, the meanings in our word problems get briefly turned over to the syntax and rules of algebra—such as setting up and simplifying expressions and solving equations—to great advantage.

In the previous chapter, we talked about how Viète saw that a wide variety of problems deep down were really problems about algebra, just cloaked in a different form. On a much smaller scale, this insight has been on display in this chapter: Our distilling of a word problem down to its quantitative essence—into the form of a variable expression and an equation—can be looked upon as an uncloaking of the problem to its bare algebraic form. Once placed in this form, it is as if the problem loses the weight and ambiguity of language and becomes decisively more transparent to handle. Algebra's development of this process to almost an act of routine exhibits symbolic maneuver at its first class finest.

Finally, let's revisit a statement made in Chapter 1:

Languages, in general, give us this wide-ranging ability to describe lots of objects and ideas with a relatively small glossary of words. Taking these words, then, in combination to form sentences—language expressions—gives us the breathtaking ability to describe nearly everything that we experience in life or are able to think about in the world around us. We seek the same in the world of numerical variations.

Hopefully, this chapter has given a small glimpse of this lofty ambition in practice. Basically, through two fundamental types of expressions (our "small glossary") in the form of multiples of x (such as $3x$, $7x$, $5x$, $55x$, $76x$, etc.), basic real numbers (such as 10, 40, 120, 3475, 17252, etc.), and their combinations, we can handle all of the variety of word problems presented in this chapter and infinitely many more problems of a similar constitution. Though the picture isn't close to being complete—due to the existence of other, more complicated variations that can't be described by the simple ones we have discussed in this book—hopefully the idea of what mathematicians want to achieve, in respect to this ambition, is a little bit clearer based on our discussions here.

Now the time has come to briefly switch gears and take a look into the external environment surrounding algebra. Throughout the first five chapters of the book, we have conceptually explored in some detail fundamental topics internal to the subject, and sought to establish connections, where possible, between them and other powerful areas of human activity and expression. Next, we turn our attention to algebra in education, an area that historically has been the source of much emotional fire and passionate commentary. In Movement 3, we take a brief journey into the fascinating world of this highly important ecosystem.

Motions in Education

The solution which I am urging, is to eradicate the fatal disconnection of subjects which kills the vitality of our modern curriculum. There is only one subject-matter for education, and that is Life in all its manifestations.

—Alfred North Whitehead (1861–1947),
The Aims of Education and Other Essays

6

The Grand Play

By whatever means it is accomplished, the prime business of a play is to arouse the passions of its audience so that by the route of passion may be opened up new relationships between a man and men, and between men and Man. Drama is akin to the other inventions of man in that it ought to help us to know more, and not merely to spend our feelings.

—Arthur Miller (1915–2005), Introduction to *Collected Plays*

To acquire an intellectual advantage at great cost, if it can be attained more cheaply is unnatural...

—Carl von Clausewitz (1780–1831), *Clausewitz and the State*

On Monday, January 6, 1930, the great American philosopher and education theorist John Dewey was invited to give a series of public lectures on philosophy and psychology at Harvard University.[1] He accepted on January 13, and one year later, he gave a sequence of ten talks so extraordinary that their content was later collected in a book called *Art as Experience* (1934).[2]

Scattered among the book's sometimes dense passages are conceptual gems of such dimension, creativity, and zing that they continue to rouse the imagination:

Every art communicates because it expresses....For communication is not announcing things....Communication is the process of creating participation, of making common what had been isolated and singular.[3]

A primary task is thus imposed upon one who undertakes to write upon the philosophy of the fine arts. This task is to restore continuity between the refined and intensified forms of experience that are works of art and the everyday events, doings, and sufferings that are universally recognized to constitute experience.[4]

We have an experience when the material experienced runs its course to fulfillment....A piece of work is finished in a way that is satisfactory...is so rounded out that its close is a consummation and not a cessation....The experience itself has a satisfying emotional quality because it possesses internal integration and fulfillment reached through ordered and organized movement.[5]

Dewey advances the view that many aspects of everyday life have the potential to be aesthetic experiences, and that they should be recognized as having this capacity. He felt that artistic activity is not exclusively the domain of the fine arts, where it is intentionally created to be experienced—as enlightenment captured or tamed—often in museums or galleries, but rather that the aesthetic experience can also happen as events unfold and unite in wild, free-flowing forms.

The artistic experience comprises a continuum between its fixed forms on the one hand and the high points of everyday life on the other, where actions and ideas converge together in unison: the smooth execution of a well-crafted, victorious game plan or campaign strategy, the successful, mature handling of a stressful task, the beautiful rendition of a song at graduation, the enjoyment of a long-desired vacation, or even a crossword puzzle or sudoku grid completed in a creative way after some struggle. Such peak experiences also include the epiphanies that occur in education: namely, moments of insight where new concepts and old memories converge together in students so that they suddenly see or understand something in a strikingly fresh or very clear way.

Dewey thought that instructors should leverage such moments in their teaching and deliberately engineer scenarios to assist students in experiencing the excitement and satisfaction of substantive understanding along with the joy of insight, which has been described as being "a sense of involvement and awe, the elated state of mind that you achieve when you have grasped some essential point; it is akin to what you feel on top of a mountain after a hard climb or when you hear a great work of music."[6]

To date Dewey is still recognized by many as the preeminent education thinker of American origin.

CONCEPTUAL FUELS

The great acting teacher Stella Adler shared some of Dewey's sentiments about unity and continuity in her own field, telling her students, "Your curse is that you have chosen a form that requires endless study....It means you have to read, you have to observe, you have to think, so that when you turn your imagination on, it has the fuel to do its job."[7]

Widely considered to be one of the twentieth century's leading teachers of drama, Adler believed that imagination and research rather than personal memories or emotional recall should inform an actor's craft, and that much of what an actor learns, observes, and thinks about can serve as fuel for their performances. It is a powerful statement because she is not suggesting passivity in having this happen either, but rather appears to advocate that actors actively and aggressively employ their individual point of view—their observations, thoughts, and imagination—as conceptual fuel for acting.

Quotations serve as fuels for the imagination, too. They are very popular, and many people are known to inventory collections of them for study and use as well. Why do they do this? Some do it for fun, but I dare say that a large portion also do it for the down-the-road purpose of later using the quotations to motivate, inspire, or enlighten others (or their future selves) for purposes or situations that may have been unknown to them at the time that they first decided to save the quote.

The thing that makes quotes so appealing to us is due in part to the fact that situations in life rhyme and sometimes an apt demonstration of a relationship or a phenomenon in another area (or at a different time) may be encapsulated in a quotation. The statement may provide a spot-on conceptual perspective or game-changing orientation to a more abstract, unpredictable, or confusing situation. Quotations communicate, across the ages, other people's singular moments and insights—serendipitous times where they have been gifted with the ability and vision to snatch something transcendent and eternal from a profound, imaginative, or confluent instant in time or thought.

When such a statement powerfully resonates with a person, it is almost like a window in time opens up for them to a wider world of shared experiences, allowing them to momentarily glimpse, "in the palm of their hand," some portion of the past, present, and future all at once.[8] These junctures become like individual little packets of Dewey-type experiences in and of themselves. And their preservation is a spectacular thing to have available at our fingertips. People's use of quotations as fuel for their own interpersonal interactions and private inspirations approximates what Adler is telling her acting students to do in employing the results of their "endless study" to fuel their various performances. It also approximates the conceptual fuel that algebra can supply in providing greater insight into certain types of quantitative situations.

WHAT TO DO ABOUT WORD PROBLEMS

Dewey's and Adler's thoughts are highly relevant to the case for algebra in general, and algebraic word problems in particular.

How best to incorporate such problems into the classroom is one of the central issues in the teaching of algebra. But it is also one of the most contentious. Most educators agree that word problems should be included in any such teaching, but they disagree mightily on how much they should be used, where they should be introduced, how they should be integrated, and what types of problems should be used.

One common complaint is that most word problems are contrived and artificial, mostly about irrelevant things or fanciful situations. Who cares anyway about silly rope lengths, perimeters of hypothetical geometric figures, the measures of made-up angles, or the number of hamburger meals in an imagined business? Is anyone ever really going to ask us to pay for something with so many quarters and dimes, exclusively?

Many critics claim that such trifles are among the reasons that so many students get bored with the subject and wind up ultimately disliking algebra. If students could only be exposed to the real uses of algebra, the thinking goes—if they could see algebra applied to realistic situations in finance, physics, accounting, biology, and computers—then they would learn to appreciate its power and relevance to the world at large.

Many experts agree with this assessment on the state of word problems in math education.

But is this really enough on its own: to simply make word problems more about realistic situations and less contrived, as in the last chapter, and then magic will happen, with students turning on to algebra and appreciating it? Noted University of Virginia cognitive psychologist Daniel Willingham doesn't think the solution is so straightforward.

In response to the 2013 *New York Times* editorial "Who Says Math Has to Be Boring?", which reiterates that real-world application is the key, Willingham writes: "So the proffered solution is real-world application. But I think a worse problem is not understanding *how* math works, being asked to execute algorithms with no understanding of what is really happening."[9] He goes on to mention in his book, *Why Don't Students Like School*, that students can be easily bored by information, even if the material is relevant to them.[10] Most teachers would probably agree with this observation.

Willingham cites his own experience at how even he can be turned off by presentations on topics critically related to his own interests and area of expertise, especially if poorly presented.[11] Many of us have probably had a similar experience of being excited and highly motivated to attend an upcoming talk, read a new book in our area of interest, or watch a film adaptation of a book series we love only to be disappointed, bored, or downright frustrated by the nature of its presentation.

All of this suggests that simply shifting the focus to real-world and "relevant" problems won't be enough to make magic happen on its own. More is required—something akin, perhaps, to sharing with students more of the types of fulfilling experiences and unity of ideas that Dewey and Adler talk about.

But where does that leave the role of word problems in math education? Are they obsolete architecture left over from a pre-digital world? Or are they an essential tool for appreciating algebra's wide-reaching usefulness? I believe that there is a place in the teaching of algebra for both contrived problems and the problems drawn from real-world applications: in the workplace, from newsworthy events, from science, or in other areas of everyday life. But in both cases, the manner of their presentation and coordination along with an awareness of the needs of their intended audience are absolutely essential to the success of the endeavor.

CLASSROOM WORD PROBLEMS

The distinction between contrived word problems and more realistic problems can be blurry at times. For instance, the break-even point example from Chapter 4 involves a scenario from business (hence real-world in appearance), yet the numbers we used were engineered such that the break-even point would work out to be a nice whole number. Moreover, they were chosen without the market research one should normally do when starting a business. All of this means that this particular problem has a real-world aspect combined with a manufactured portion.

In recognition of this, we will make a slight adjustment in our classifications. Now, in addition to the truly manufactured situations such as the number of days and age problem (or rope length problems), we will join together those real-world-flavored applications and intentionally sculpted scenarios not likely to actually occur—and place them under a new category that we will call "word problems for the algebraic classroom" or, more simply, *classroom word problems.*

For now, we will reserve the category of real-world applications for those quantitative problems that some person might naturally encounter outside of an algebra class or recreational math book. There will obviously still be much overlap between these two redefined categories, but these are the rough subdivisions that will be meant in the discussion that follows. I share Willingham's view that we should teach for understanding, a deeper type of understanding that involves more than simply knowing how to execute algorithms. This understanding certainly includes knowing how to process algorithms, but it also includes knowing *why* those methods work—how they can organize, clarify, and generalize information to a wider world of possibilities. In other words, we must try to give students better and more representative samplings of the comprehensive understanding that an expert has.

This is a noble goal, which undoubtedly the majority of math educators share. But how do we get this to really happen?

The most common method, historically, aimed to do this by immersing the student in a rich bath of procedural know-how splashed with mostly classroom word problems of various types—not teaching specifically or long enough to achieve that deeper holistic understanding, but hoping instead that with sufficient exposure to the procedures (through repetition and taking enough classes), most students would eventually acquire it. But in the mind of some well-respected authorities, aiming to produce genuine understanding this way is a "forlorn hope."[12]

This historically common method has always been and continues to be the target of some of the harshest sustained criticism leveled at math education, which has spawned many large-scale reform efforts throughout the last two centuries to improve upon it or do away with it entirely. These efforts, from both inside and outside of the profession, are still ongoing.

Teaching the subject to students this way, however, has not been the unmitigated disaster that it has sometimes been portrayed to be. In fact, many students actually do learn a good amount of algebra and often become quite proficient with the algorithms—certainly more now than a majority of their counterparts two or three centuries ago: some to the point that they actually end up liking and appreciating the subject and going on to take more courses.

Nevertheless, the method has real issues if so many students leave the classroom with a healthy resentment for the subject rather than an appreciation of what it can do. Even for those students who like algebra, success in class does not necessarily translate into real meaning outside of the classroom for many of them. People like and play board games well too, but in most instances such games have no meaning outside of the context in which they are played.[13]

Moreover, far more disconcerting is the fact that not only does such teaching not get most students to the levels of insight that we want, but it also leaves a larger portion of them with the diametrically opposite view of mathematics than we want them to have—which is viewing the subject as an incomprehensible and meaningless exercise in symbolic manipulation. Something with little relevance to them. Something to be feared and avoided.

All of this suggests that simply exposing people to topics and having them perform exercises is not enough with algebra. In algebra a student can hear an instructor talk about the subject and see them work multiple dozens of examples in front of them, then can participate by doing many more problems themselves (ad nauseam), yet may never experience (on their own) the levels of usable understanding, comfort, and appreciation that we want them to obtain.

THE SUNLIGHT OF WORD PROBLEMS

If we want students to experience the satisfaction of a deeper and more holistic understanding of algebra, the evidence suggests that we have to explicitly teach for that type of understanding up front. Hoping that it will arrive by osmosis for students only through the exposure of having to take required classes (which occasionally hint at it) just does not work for the vast majority of people. In short, if we want students to experience the wonder, satisfaction, and appreciation of algebra, then we should probably put these into our teaching right from the start and make it transparent to them throughout.

However, there are still no guarantees. Even if we are able to find better ways to teach for deep understanding, it is still a tall order to expect that the majority of students will reach the levels of appreciation and comprehension that we hope for in the brevity of time available. Simply put, effective public education for all—even in a perfect world—seems to be a genuinely complicated problem, especially in mathematics where learning abstract and unfamiliar content and norms is hard.

Educators, past and present, are not simply making these difficulties up.

The enormity of this task is further magnified by the fact that some students simply don't care, have insufficient preparation or bad study habits, often forget what they've learned (regardless of age), experience severe math anxiety, test poorly, and are many times also confronting (on a daily basis) the extremely serious issues of life, environment, and home. Adding to this complex soup is the fact that methods that work for one group (e.g., returning adult students) may dramatically founder when used to instruct another group (e.g., ninth grade students).

The pervasive presence of these critical realities should be recognized and appreciated by any who would weigh in on math education—especially those with the ability to influence policy or outcomes.

But the potential for using word problems to aid in comprehension and appreciation is substantial, and excellent instructors have tapped this well throughout the decades. Much of this potential can be reached through the use of classroom word problems. Let's discuss some of the direct benefits of using these types of problems:

- Such word problems allow us to work with simpler numbers, meaning not only small values but also round or whole numbers, integers, and fractions. This is a distinct advantage over many real-world applications, which can often involve more complicated fractions and decimals and require more initial maneuvers to translate into the language of algebra. More complicated situations can distract students, causing them to focus on other computational and translation issues rather than on the algebraic structure of the problem, which is where we usually want them to be.[14]
- They generally require less background knowledge than true real-world applications. Problems from science and business often require students to know or be introduced to a fair bit of information about the fundamentals of physics, chemistry, computer science, finance, accounting, or biology before they can understand the essentials in a problem. As students try to synthesize information across multiple fields of study, this can be an obstacle to the primary function of the problem—learning algebra.
- They showcase that it is possible to take a problem in language—often containing inconvenient or difficult-to-obtain information—and drill it down to its quantitative essence, thus allowing us to uncover the key algebraic relationships involved, from which it becomes possible to obtain answers in a systematic, almost recipe-like fashion. Many classroom word problems naturally illustrate this near feat of magic: that of turning mathematical sophistication into routine.

- They allow for the assembly-line production of situations based on many possible variations of a single conceptual theme. This allows students to practice the same algebraic ideas in different contexts as they learn to set up the algebra, simplify expressions, and solve equations. Real-world problems often have a custom-made feel to them that, if exclusively relied on or improperly coordinated, can make it hard for algebraic novices to gain a sufficient enough foothold to acquire the necessary understanding and confidence in the subject for effective appreciation and use. Such production helps us explore how tweaking the algebraic expressions and parameters can give us insight into new situations.

On the whole, because of these properties, classroom word problems can give students a more unified experience in algebra, showing them that the subject is a thought-provoking, highly interconnected, and, dare I say, beautiful entity in its own right.

THE MOONLIGHT OF WORD PROBLEMS

Other, more subtle benefits can also be obtained from the use of classroom word problems. Embedded within such problems are realistic and deep properties of math and nature that can be teased out.

The math subcommittee to the influential *Committee on Secondary School Studies* (1890s) had something to say on the matter, and though the following was written specifically in their comments on arithmetic, it certainly applies to algebra as well:

> The pupil who solves a difficult problem in brokerage may have the pleasant consciousness of having overcome a difficulty, but he cannot feel that he is mentally improved by the efforts he has made. To attain this end he must feel at every step that he has a new command of principles to be applied to future problems. This end can be best gained by comparatively easy problems, involving interesting combinations of ideas.[15]

I doubt that the educators on this subcommittee, such as Florian Cajori, Simon Newcomb, and William Byerly, were explicitly opposed to teaching students how to use mathematics to solve a customized brokerage problem; rather, they were simply more interested in stressing the potential power for explanation—and the demonstration of method—inherent in the judicious employment of basic problems. In other words, they were advocating that "smart campaigns for deep understanding" (that include a student's attitudes and impressions) be put in place when using problems in the mathematics classroom.

This thinking around using classroom word problems is similar in spirit to how physicists teach the motion of falling objects in basic classes. Although the drag created by air resistance is a very real effect that must be accounted for to obtain accurate calculations on Earth, they often first ignore this in instruction, as including it can greatly increase the level of mathematical difficulty for most students—to the point of crippling distraction.

The trade-off is that the answer students obtain with these "training wheels" on in effect isn't as precise as it could be, but that isn't always the goal of an exercise in the introductory physics classroom. It may sound counterintuitive that the point of solving a problem in this context is not to obtain the most correct solution, and certainly a working physicist would need to prioritize accuracy over simplicity. In education, however, this type of simplification enables students to focus on learning fundamental aspects of motion and forces, and how gravity, accelerations, velocities, heights, and time interrelate, as well as how to systematically use mathematics to deal with such phenomena. The more exact details can be dealt with at a place further along in the student's education.

This thinking also shares similarities with the way scientists acquire fundamental facts about nature through experimentation. The setups of many of these experiments are often highly choreographed and artificial. And being seen as deliberate staging, such approaches have had their critics—the philosopher Thomas Hobbes among them.[16] But because the laws of nature don't turn off in these cases, even staged and carefully controlled scenarios can be useful tools for illuminating such laws.

Indeed, it is precisely the fact that the laws still work in such cases that allows scientists to isolate a particular property or aspect of a phenomenon from the others in an experiment so as to obtain deep insight. The last 400+ years have shown us the overwhelming potency of employing this investigative method to learn about nature (and this includes Galileo's and Einstein's highly productive thought experiments, too).

In much the same way, classroom word problems being used in close coordination can allow us to focus in on various structural things in algebra. It is as if we can make up lots of different little algebraic experiments to shine a powerful spotlight onto some of the deep, fundamental properties of quantitative variation. But these properties must be explicitly pointed out to students, as they generally won't recognize them on their own; algebra is extremely clever at masking its tracks.

Let's take a look at a few of the less obvious properties that popped up during our use of classroom word problems:

- Consider the "variable expressions" viewpoint versus the "unknown values" viewpoint: The "variable expressions" viewpoint illustrated that although we initially thought we were solving only one specific word problem, the problem actually turned out to be just one instance of an entire family of rhyming word problems (see Word Problem 3 in Chapter 5). More generally, this viewpoint provides an avenue to a smoother transition to the function concept, which becomes so important elsewhere in mathematics and in the physical sciences.

  The "unknown values" viewpoint helps us to see that although we may not know the value of a given number, we can still incorporate it into expressions and manipulate it—as if we did know the number—thus projecting our knowledge, in some cases, far beyond what we knew at the start of the process.

- Drilling down to the algebraic essence of problems showed us that many situations, though dressed up differently, have quantitative similarities. The problems in the last chapter involved unknown measures of angles, ages, different types of coins and bills, and rope lengths. Yet when we distilled them down to their quantitative essence, we saw that they were all solvable by highly similar algebraic expressions and equations.

- Simplifying algebraic expressions and equations shows how variation and stability collectively mix together. We may have several quantities of assorted types in a single problem that are changing together simultaneously, and in our quest to understand what is going on, natural questions arise about the net result of their interaction.

  The basic rules of algebra, which allow us to tag different types of variation and combine together those of the same type, can then be employed to significant effect in giving us the ability to better understand and sort out this collective behavior (see "Separating Out Numerical Interactions" in Chapter 2).
- Algebra strengthens arithmetic skills. Practice makes more perfect, and in order to do algebra, one must necessarily do the arithmetic embedded in the process. Many students who enter an algebra course struggling with their numerical skills—especially in regard to integers and fractions—emerge on much more solid computational ground, even if they still haven't fully grasped the algebra itself. Put another way, we don't stop learning arithmetic simply because we have started studying algebra.
- We can get a lot of bang from very simple combinations of variations represented only by multiples of x ($6x$, $55x$, etc.) and basic numbers (-5, 10, 17, etc.). Word problems come in infinitely many faces and forms, yet multitudes of them can be described through the use of basic terms involving only numbers and multiples of the simplest variation represented by x (such as $5x + 20 = 180$ or $76x = 17252$).

Though such descriptions are certainly not exhaustive, there are a wealth of situations involving more complicated variations like x^2 or 10^x—the range of what can be produced from mixtures of just these two simple components is amazing.

CONCLUSION

Throughout this book, we have explored a wide assortment of problems, situations, concepts, metaphors, and story lines. Yet out of that medley, a "grand play" has slowly emerged and coalesced into something that we can now characterize as a bona fide algebraic experience, including the following:

Variation: Observing a numerical ensemble in a multitude of guises.
  ◦ A central procedure or theme that produces diverse numbers for different individuals, organizations, times, locations, and so on.
  ◦ An unknown piece of quantitative information that satisfies certain conditions.
Symbolic Representation: Capturing the essence of the variation or conditions with an algebraic expression or equation.
Maneuvers: Simplifying the expression or solving the equation to gain insight.
Analysis: Interpreting the results of the representations and subsequent maneuvers.

Though this summary captures key points in the process, it doesn't reveal the entire scope of what most educators would like this basic algebraic experience to also consist of.

What does this really mean? Imagine that you're planning a vacation to the American Southwest. What are the elements that will make your vacation a true vacation? Some might fixate on details like making reservations for hotels and restaurants, packing the car, and visiting national parks. Others might point to the logistics of how travel happens, like using a GPS, keeping the cooler stocked, and making regular stops for gas. However, though these all are definitely aspects of vacationing, something is missing if we only define a vacation by these terms alone. Doing so would omit the qualitative experience of traveling and your motivation for doing so in the first place, be it rejuvenation, recreation, the desire to experience something new, or witnessing first-hand the aesthetic beauty of the spectacular scenery in canyon country.

So it is with algebra. Though the four components listed earlier are essential to the algebraic experience, they don't singularly convey the reasons why most mathematics professionals want to keep algebra in the curriculum, which include teaching students

- to see how basic algebraic thinking and expressions can help clarify what's really going on in confusing or abstract quantitative situations,
- to see how the algebraic way offers insight into numerically varying phenomena in general,
- to gain technical confidence and comfort with algebra as a tool for understanding,
- to make illuminating connections that generate excitement and gratification in understanding something new,
- to genuinely appreciate the scale and beauty of algebra and mathematics, and their connections to the world at large,
- to acquire the necessary skills for further exploration of mathematics in other classes or on their own.

It is quite remarkable that a huge chunk of this can probably be accomplished through just the use of carefully coordinated classroom word problems and situations alone—with the accompaniment of explicit and guided instruction on what is happening. And, if so possible, why not use such problems to do much of the heavy lifting, as the quote attributed to Clausewitz figuratively suggests?[17]

The internal consistency and interconnectedness of classroom word problems, however, can't be all there is to it. If so, then this algebraic experience—as neat as it can be in cloistered form—would still probably not be deserving of the central place that many educators want it to hold as a required subject in the K–14 mathematics curriculum.

Still more is expected and demanded of a subject that is mandated to touch armies of students—meaning that at some point this experience must be shown to be somewhat relevant in areas that we might actually encounter outside of the classroom. Think of the internal workings of a modern flat-screen television. The coordination of millions of thin-film transistors combined with light and color technology is truly a wonder. But if all these devices were useful for was the wonderful things happening on the inside, most of us, outside of some scientists and engineers, would give nary a thought to them.

The thing that makes tens of millions of us care about flat-screen televisions (and want to bring them into our homes in multiples) is the fact that the successful operation of their internal workings opens wide to us the vast, complicated, and diverse world out there—offering instantaneous, high-resolution access to images from all over the globe as well as entertaining dramas, news, sports, commentary, documentaries, and so on that, taken together, have had an enormous impact on our lives.

And while algebra may never be as popular as television, many of its concepts and techniques are not confined to the inside of the classroom, either. They live outside of it in powerful, and often subtle, ways that have dramatically changed the collective lives of human beings, for better and for worse. Moreover, the methods exhibited in this "grand play" offer us the chance for better awareness and insight into the steady stream of quantitative information that we face on a daily basis—if we only tune in more to what they can tell us, that is.

In the next few chapters, we'll tune the algebraic antennae a bit more to better listen in on some of the quantitative variations out there—variations that we may personally encounter or hear about in the course of our daily lives. But first, we explore a bit of the history of the quest to introduce algebra into the public school curriculum and the ensuing efforts to keep it there.

7

Algebraic Awareness

The idea of requiring algebra for all remains under assault from many
corners in the United States today. Burning like an eternal flame, still,
is the centuries-old argument that most people won't ever use the algebra that they learn in school anyway, so why make them take it at all?

Such thinking contrasts sharply with the heady days of the late
nineteenth century when influential educators, no doubt in part after having observed the spectacular successes of injecting arithmetic
earlier into and throughout the curriculum, wanted to see something
similar happen on a wider scale with algebra. At least in the public high
schools, that is, whose day in America had arrived.

Going back a hundred years further still to the 1700s, even something as simple and practical as basic written arithmetic was not in
the public trust at all, generally being taught only to a relative hand-
ful of privileged students of adolescent age. This practice had its roots
going all the way back to thirteenth-century medieval Italy with the

introduction of the then revolutionary abbaco schools, which taught select students in this age group written arithmetic (among other subjects) for commerce—in the vernacular instead of Latin.[1]

The idea of an organized public education for all was still in its infancy in the eighteenth century, with most young Americans not attending school at all. And for those who did, most ended their formal education long before their adolescence. Thus, most Americans knew very little arithmetic. Such was the state of math education in this country that arithmetic was, for a time, taught in the junior and senior years at Harvard College. It was only sometime around 1780 that the subject was moved down to the freshman year.[2,3]

George Washington thought it an issue of patriotic concern. In his 1788 letter to Nicholas Pike, author of the first arithmetic textbook written by an American after the revolution (*A New and Complete System of Arithmetic Composed for the Use of the Citizens of the United States*), he states:

> But I should do violence to my own feelings, if I suppressed an acknowledgement of the belief that the work itself is calculated to be equally useful and honorable to the United States.... The science of figures, to a certain degree, is not only indispensably requisite in every walk of civilised life, but the investigation of mathematical truths accustoms the mind to method and correctness in reason.... From the high ground of Mathematical and Philosophical demonstration, we are insensibly led to far nobler speculations and sublime meditations.[4]

Elsewhere in the world at the time—notably in Switzerland—remarkable educational reforms were being proposed and developed. And one of the most significant curricular innovations to emerge from it all was to be elementary arithmetic. The reforms in its teaching and presentation would be of such a force, excitement, and extent that important parts of the subject would soon storm their way spectacularly down into the public elementary school—eventually coming to rival even reading, vocabulary, writing, and spelling, themselves, in importance.

Arithmetic's stunning migration down to the earliest years of the public grade school (wresting it out of the hands of the privileged and the college-educated, thus making it potentially accessible and understandable to nearly everyone), though not perfect, still remains one of the most successful and enduring educational reforms of all time. We are all among the beneficiaries of this late eighteenth- and early nineteenth-century reform in arithmetic, a reform so successful that few still seriously question the subject's place in education. It has become one of the pillars in elementary instruction.

Nothing close to similar has happened, unfortunately, with the introduction of algebra into the American public school curriculum. The educational reform cementing its near universal comprehension, acceptance, and appreciation—though highly desired by many in the late nineteenth century—still has yet to occur.

Why is this? An enduring educational puzzle? Or does algebra just have a PR problem?

LOFTY, ILLUSORY GOALS

The goals for the inclusion of algebra in the American public school curriculum have always been lofty and somewhat elusive. The subject in many ways took a parallel trajectory to that of basic arithmetic, as both were commonly taught at the college level early on before being pushed down a level or two in the curriculum. Its retention, along with geometry, as part of the standard college course of study was vigorously defended and leveraged in the highly influential "Yale Report of 1828."[5]

The push for injecting algebra into the secondary school curriculum can be seen in the gradual incorporation of the subject in some of the private academy (private secondary or high school) textbooks of the eighteenth and early nineteenth centuries.[6] Perhaps this was done initially to give the more advanced students an advantage before college—as AP courses do today—but by 1820, Harvard had made it a requirement for admission, spectacularly going in just over four decades from teaching its students arithmetic in the upper class years to now requiring that they knew not only arithmetic but also some algebra before even being admitted to the institution.[7]

The relatively small but growing number of public secondary schools also began to incorporate algebra into their curricula, and by 1850 its place in the high school was no longer out of the ordinary.[8] The inclusion of other less traditional school courses for the time (history, geology, art, music, bookkeeping, surveying, physical education, modern languages, training for future teachers, modern literature, geography, chemistry, and manual training) followed suit, leading to a veritable explosion in the diversity of curricula around the country.[9] By the latter part of the century, inconsistent course offerings from school to school and competing philosophies about the goals and purposes of secondary education—indeed all of public education—were becoming a burden for colleges and universities to decide on what to accept for admission.

And even more strongly in turn, the diversity in admissions requirements and course offerings from the many colleges and universities along with an increasingly diverse student body were beginning to overwhelm some high schools on which subjects and prerequisites (if any) they should fit their curricula to, because it was becoming more and more difficult to satisfy the needs of everyone.[10]

Noted educational historian Diane Ravitch summarizes the issues of the day nicely:

> In an age marked by the development of systems and organization, the schools seemed helter-skelter, lacking uniformity or standards. What should be taught? To whom? At what age? For how long? What were the best methods? What subjects should be required for college entrance? Should "modern" subjects such as history and science be accepted for college admission? Should students be admitted to college who had not studied the ancient languages? Should there be different treatment, even different curricula, for the great majority of students who were not college-bound? Should high schools offer manual training and commercial subjects?[11]

In a sweeping effort to provide some professional guidance, the National Education Association (NEA) in July of 1892 formed a committee of ten of the most influential educators in the country to study the problem and lead the charge to create a set of standards and guidelines for the American high school curriculum.[12]

Though the entire project was officially named the Committee on Secondary School Studies, the collective effort came to be better known as the famous "Report of the Committee of Ten"—so named for the ten-man committee that facilitated the project and composed the final report. However, in compiling the final report, the committee was advised by nine subcommittees (each with ten members) to ensure that their recommendations were informed by subject-matter experts and educators, one for each of the following subject areas[13]:

1. Latin
2. Greek
3. English
4. Other Modern Languages
5. Mathematics
6. Physics, Astronomy, and Chemistry
7. Natural History (Biology, including Botany, Zoology, and Physiology)
8. History, Civil Government, and Political Economy
9. Geography (Physical Geography, Geology, and Meteorology)

It is worth pointing out that, at this time, higher education itself was also in a state of tremendous flux, with the idea of the classical college and its prescribed curriculum as endorsed by the "Yale Report" now giving way to that of the modern university with its elective curriculum and postgraduate and professional schools. Thus, the Committee of Ten and its subcommittee members—47 from colleges/universities, 42 from schools, and one a government official who formerly worked at a university—were in the vanguard of these seismic changes and among the most reform-minded educators in the country.[14]

The subcommittees each met in three-day-long conferences in late December 1892. NEA Board Chairman (and noted education author) Norman A. Calkins had attempted to negotiate with the railroad companies to get reduced fares for conference attendees who were traveling from afar, but though able to procure some reductions, he was not as successful as had been hoped.[15] The conference locations and dates were as follows[16]:

Conference	Host Institution	Location	Dates
Latin	University of Michigan	Ann Arbor, MI	December 28–30, 1892
Greek	University of Michigan	Ann Arbor, MI	December 28–30, 1892
English	Vassar College	Poughkeepsie, NY	December 28–30, 1892
Other Modern Languages	Bureau of Education	Washington, DC	December 28–30, 1892
Mathematics	Harvard University	Cambridge, MA	December 28–30, 1892
Physics, Astronomy, and Chemistry	University of Chicago	Chicago, IL	December 28–30, 1892
Natural History (Biology including Botany, Zoology, and Physiology)	University of Chicago	Chicago, IL	December 28–30, 1892
History, Civil Government, and Political Economy	University of Wisconsin	Madison, WI	December 28–30, 1892
Geography (Physical Geography, Geology, and Meteorology)	Cook County Normal School	Englewood, IL	December 28–30, 1892

The discussions at the meetings reportedly were "frank, earnest, and thorough"; yet in the end, they resulted in remarkable agreement amongst the various groups (with only two subcommittees submitting minority reports). The final reports from all of the conferences were completed by the middle of July 1893.[17]

Each of these reports were delivered to the Committee of Ten, and by December 1893, they were ready to share their final consolidated report. This report is still, over 125 years later, one of the most influential educational documents ever issued in the United States.

Some of the harshest criticism against requiring algebra, among other courses, in the high school curriculum today is often directly or implicitly aimed at the report's conclusions and the individual members of the Committee of Ten. Unfortunately, much of it, according to

Ravitch, David Angus, Jeffrey Mirel, and other historians of education, is based on myths, half-truths, and outright mischaracterizations on the committee's aims—as well as on misconceptions of the state of American high school education at the turn of the century.[18] Much of it also seems to ignore the fact that the case against algebra isn't new. Powerful, detailed, sustained, and successful crusades have already been waged in the past against requiring algebra and other academic subjects in the secondary curriculum—especially during the first part of the twentieth century—with highly problematic long-term effects.[19]

On the eve of Pearl Harbor in November of 1941, Admiral Chester Nimitz himself, in a widely circulated letter, expressed alarm about what he saw as the decreasing level of mathematical education in naval officer candidates. Many educators—naval and civilian—attributed this decline to the more than 25-year assault on basic mathematics instruction that had been waged by prominent leaders in the progressive education movement.[20] The alarm bells continued to sound after the war, most prominently at the University of Illinois in the early 1950s, leading to an influential reform movement, whose math component was initially spearheaded by Max Beberman, Gertrude Hendrix, and Herbert Vaughan, and that ultimately culminated in the New Math of the Sputnik era.

Even in their day, the views of the committee were controversial. Committee Chairman and Harvard president, Charles Eliot, who had spearheaded curricular reform through pushing for the elective system over the traditional prescribed curriculum in American colleges, stated in 1892:

It is a curious fact that we Americans habitually underestimate the capacity of pupils at almost every stage of education, from the primary school through the university. The expectation of attainment for the American child, or for the American college student, is much lower than the expectation of attainment for the European. This error has been very grave in its effects on American education...[21]

Florian Cajori, a prominent member of the math subcommittee, shared similar sentiments, stating two years earlier in 1890:

One of the most baneful delusions by which the minds, not only of students, but even of many teachers of mathematics in our classical colleges have been afflicted is, that mathematics can be mastered by the favored few, but lies beyond the grasp and power of the ordinary mind. This chimera has worked an untold amount of mischief in mathematical education....This humiliating opinion of the powers of the average human mind is one of the most unfortunate delusions which have ever misled the minds of American students and educators. It has prevailed among us from the earliest times.[22]

Such pronouncements give some indication of why the committee wanted to keep subjects like algebra in the high school curriculum.

Critics argue that the Committee of Ten members—mostly of privileged background—were out of touch with the needs of the new crop of students entering high school, and that their elitist outlook biased them toward a view of public high school solely as preparation for college-bound students with no thought or concern for students whose education would end with secondary school.[23] However, the 1893 report itself states:

The secondary schools of the United States, taken as a whole, do not exist for the purpose of preparing boys and girls for colleges. Only an insignificant percentage of the graduates of these schools go to colleges or scientific schools. Their main function is to prepare for the duties of life that small proportion of all the children in the country—a proportion small in number, but very important to the welfare of the nation— who show themselves able to profit by an education prolonged to the eighteenth year, and whose parents are able to support them while they remain so long at school....A secondary school programme intended for national use must therefore be made for those children whose education is not to be pursued beyond the secondary school. The preparation of a few pupils for college or scientific school should in the ordinary secondary school be the incidental, and not the principal object.[24]

This is an extremely ambitious program for the secondary school, then sometimes called the "People's College". At its root were two be-

liefs: (1) A broad-based education in some combination of the nine subject areas was the best way to develop a well-informed and reasoned
American individual, and (2) all high school students, college-bound
or not, should have equal access to such studies once considered open
only to the most privileged of youths. This course of study was deemed
by committee members to be important for participatory citizenship
and leadership in the democratic United States, regardless of eventual
employment destination.[25]

Consider literacy, and how reading is seen as a vital core skill required
to navigate in the modern world, no matter what else a student may
learn or do in the future. To the Committee of Ten, certain parts in the
nine subject areas were also important for a citizen, of the time, receiving a high school experience or certification to be proficient in—not college as an end in itself, but a solid broad exposure to what they viewed as
being among the central and rhythmic strands of knowledge. This was
most especially true for those students who would not continue beyond
high school, because committee members felt that (for these students)
this exposure in secondary school would be their only chance to ever
explore these strands in a systematic way before moving on to receive
more job-specific training in their chosen occupations. Afterward, college could also be at the ready if some of these students later changed
their mind about continuing their education. This approximates an essential tenet in the majority viewpoint of the Committee of Ten.

Even by 1893, the arguments for continuing to accord classical languages like Latin and Greek such prominence in the curriculum were
steadily losing ground, as a prime justification for them, as sharpeners
of the mind—the mental discipline theory—increasingly came under
attack. Additionally, as more comparatively recent literature such as
Shakespeare's works and the works of American authors gained increasing acceptance into the curriculum, and as translations of works
in other languages became more available, many believed there was a
decreased need for the ancient languages to be required any longer for
the reading of ancient literature. Thus, it became ever more difficult
to continue selling the idea that the study of these languages offered
any marked advantage over the study of modern languages. Evidence
of this can be seen in the steady decline in prominence of Greek and

later Latin in the secondary curriculum, which continued on into the twentieth century.

The seven other subject areas had a more robust array of defenses to bolster up their claims as to why they should be a part of the core strands of general knowledge.

William Torrey Harris, US Commissioner of Education and a member of the Committee of Ten, had stated many years earlier that a standardized curriculum for public education was necessary in order to avoid creating a caste system in America, where such education would be reserved for students of privilege in private academies and not available to students who attended public schools.[26] He viewed such a system as being fundamentally nondemocratic.[27]

Others didn't think it possible nor desirable to achieve these aims with public secondary education for all. One of the harshest opponents of the Committee's recommendations was G. Stanley Hall, president of Clark University in Massachusetts and an internationally respected psychologist. Hall later criticized the Committee's conclusions because he thought that their aims were too idealistic and that statements such as those made by contributors to the final report do not "apply to the great army of incapables, shading down to those who should be in schools for dullards or subnormal children, for whose mental development heredity decrees a slow pace and early arrest, and for whom by general consent both studies and methods must be different."[28] Hall's strongly worded opinion, encapsulated in this one hyperbolic statement, has for over a century kept his name at the forefront of those cited as being in disagreement with the Committee's conclusions.

Many felt similarly to Hall regarding the capacity (or lack thereof) of the average American student—even those who would have chosen to use less condescending language—arguing for less standardization and for the creation of significantly more varied educational tracks for high school students of "varying ability and interests." Still more conservative commentators felt that the Committee had already been far too generous in its reforms by proposing to include "newer" subjects such as biology, geography, and history and stressed that the classical, prescribed curriculum as recertified in 1828 with Latin and Greek at its core should be shored up in the high schools and adhered to—especially as the battle to retain the classical curriculum at the post-secondary level was steadily being lost.[29]

The mathematics subcommittee tried to strike a balance in its recommendations. The group recognized the enormous difficulties many students faced under the existing systems of instruction, and realized that substantial change was necessary to deal with them, arguing for the need to create a more integrated understanding and appreciation of mathematics in the student. To do this, they suggested removing the especially arduous or arcane problems that were found in some of the textbooks of the day, as well as replacing some of the more demanding and opaque arithmetic procedures with the simpler, more unifying algebraic methods.[30] These latter recommended revisions took firm root in the 1900s.

The math subcommittee also campaigned for a different type of in-class instruction, one that didn't rely on rote memorization alone, but rather greatly enhanced the powers of memory by motivating the subject first through the use of more concrete examples.[31] Additionally, they wanted to see instructors become more effective at explaining why mathematics works and how to apply it, believing that doing so would make the subject more understandable, broaden its appeal, and better accommodate the tidal wave of new students coming into public high schools.

Comparing nineteenth-century arithmetic and algebra textbooks with those from the twentieth century, we can see that implementation of some of these ideas did occur and has resulted in some positive progress.[32] However, if these arguments for reform still sound familiar today—125+ years and running—it may be because the subcommittee's lofty goals to make math more accessible and appealing haven't yet been fully realized. Most contemporary educational narratives—in the press and elsewhere—don't indicate disagreement.

Though the members of the controlling Committee of Ten in their final report were a bit firmer in their recommendations for the math course of study, the ten members of the math subcommittee had allowed for diversification beyond a certain point. But for them, that diversification (e.g., commercial mathematics and bookkeeping) should happen after the first course in algebra, not before.[33] A number of the arguments today passionately contend that this diversification should happen before the first course in algebra, not after.

The following table shows for a given date the percentages of students in the last four years of high school enrolled in algebra, geometry, and trigonometry[34]:

School Year	Algebra	Geometry	Trigonometry
1890	45.4%	21.3%	1.9%
1900	56.3%	27.4%	1.9%
1910	56.9%	30.9%	1.9%
1915	48.8%	26.5%	1.5%
1922	40.2%	22.7%	1.5%
1928	35.2%	19.8%	1.3%
1934	30.4%	17.1%	1.3%
1949	26.8%	12.8%	2.0%
1952–1953	24.6%	11.6%	1.7%
1956–1957	28.7%	13.6%	2.9%

For the snapshot taken in 1890, 45.4% of all students then enrolled in the 9th, 10th, 11th, and 12th grades were taking a course in algebra during that year.

ALGEBRA FOR WHOM?

Whatever the argument, the importance of algebra to the existence and continued expansion of critical sectors in our modern high-tech society is without question. This is even more true now than it was in 1893. For starters, many new technologies and modern conveniences in transportation, communications, energy, electronic computation, military science, and consumer products—that we've come to enjoy over the last 125 years—owe their existence to advancements in science and technology, fields that critically depend on mathematics.

So, someone needs to learn algebra. Few, if any, dispute this. But should everyone be required to learn some?

Clearly, those students who pursue advanced studies in a STEM— science, technology, engineering, and mathematics—field will use the elementary algebra they learn in school. But what about students whose sole exposure to the subject will be a single course or two? Can they really take away anything of lasting substance and value from the experience—other than way more frustration and grief than most of them should endure?

Similar questions, regarding more advanced courses, can be raised by STEM students in higher education. These students, particularly engineers, are required to take the calculus series (and often a math class or two beyond), and a lot of them struggle mightily in these classes, too. Later on, many discover, especially with the tools of modern technology, that they rarely use this higher-level math professionally in the form it was taught to them in college. So, they might just as easily ask: What exactly are they taking away from these more advanced math courses?

Yet, most engineers don't seriously ask this question. Why is this? Whether or not the more advanced math courses have a direct practical application in their working life, most engineers know that these courses give them powerful and useful conceptual fuel. We can see this reasoning manifest itself in several ways. For one, from a strictly pragmatic standpoint, most engineers know that more advanced mathematical knowledge qualifies them for a broader pool of jobs in their field. This is not an insignificant factor, as many people end up having several different jobs over the course of their lives—and as Stella Adler reminds us, the more comprehensive our education, the better our ability to perform in a variety of roles.

Another way, perhaps, to think about this is in terms of active and passive vocabulary. A person's active vocabulary consists of the words that they regularly use in their thinking, writing, and speaking, whereas their passive vocabulary includes all of the words they recognize and understand.[35] If we apply the logic that is often leveled at the algebra requirement toward our vocabularies, then a person shouldn't have a passive vocabulary at all. After all, why learn words that you aren't going to regularly use?

But clearly our passive vocabularies are important too, as they enable us to understand and appreciate a wider range of concepts than we can ourselves directly communicate. We can, in a short span of time, read a book that may have taken a seasoned expert over a decade of concentrated effort and research to write, thus becoming an active participant in a sophisticated intellectual activity that we could in no way have produced ourselves. Our passive vocabularies grant us the ability to more effectively dial in and interpret parts of the greater world outside of our own experience. This is even more apparent in young children who can create far less in language than they are able to understand and act upon—and what they hear during those critical early years of life often makes a decisive difference in the rest of their lives.

This leads into our second point. Engineering students also bring a mathematical awareness into their professions from their coursework that allows them to comfortably understand much more technical material than they regularly use or ever need produce. This awareness serves, in a way, as part of their passive technical vocabularies, without which most would not be able to perform their jobs with confidence. Mathematicians sometimes liken this mathematical awareness to a type of maturity—often calling it "mathematical maturity." Both points still retain some water if we bring them back to the level of elementary algebra, where we will call the idea "algebraic awareness."

This leads us into the chorus of this chapter and the next several chapters, the goal of which is to showcase a few of the recurring narratives and enhanced outlooks from which we can benefit by developing and tuning into this awareness.

THE STAR-SPANGLED BANNER

On a given summer day in the United States, our national anthem is sung in dozens of ballparks and sporting venues across the nation. Although each performance may be sung by different individuals, each with a different take and sound, we do not balk for a single moment in declaring that they are all singing the same song—"The Star-Spangled Banner."

What is it about each unique performance that causes us no strain in identifying them all as being the same piece of music? Clearly, the fact that we recognize the same lyrics and tune is what binds all of the renditions together for us, making us look on each of them as simply different versions or interpretations of the same song.

A few members of this mighty ensemble are listed next:

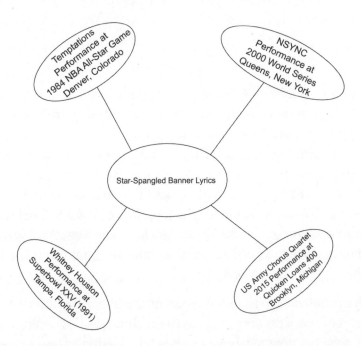

This visual—of explicitly stated common structure (the lyrics) amidst all of the variation (the different performances)—has parallels all over the mathematical landscape and its application, providing us with potent conceptual fuel.

ALGEBRAIC SONGS

Consider the following five problems:

1. An unknown number added to ten more than fifteen times itself gives one hundred six. Find the number.
2. A 106-foot length of rope is cut into three pieces. The second piece is 10 feet longer than seven times the length of the first, and the third piece is eight times the length of the first piece; find the lengths of all three pieces.

3. Given a rectangle of perimeter twice 53 meters and whose length is five more than seven times its width, find its length and width.

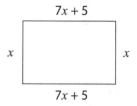

$$7x + 5$$

x ⬚ x

$$7x + 5$$

4. If a power tool rents for $16 a day plus a one-time $10 processing fee, how many days can you rent the tool if you have four $20 bills, two tens, a five, and a dollar to spend?

5. Next Saturday, Barbara will be doing a job for a client that pays her $36 an hour. The job requires specialized computer services that cost $20 an hour to use in addition to a $40 setup fee. When she arrives on location for the job, she notices a $50 bill that the client left to say thanks for coming in on the weekend. How many hours does she need to work so that her total profit (including her tip) for the day is $106?

These problems are all about different things on the surface, but setting each of them up according to the techniques discussed in Chapter 5 shows that their algebraic essence is quite similar (see Appendix 2 for their analyses). In each case, simplification ultimately yields the equation $16x + 10 = 106$:

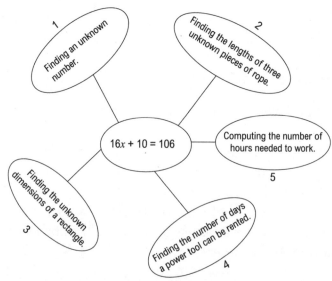

Just as we can, through the lyrics, identify the many performances of "The Star-Spangled Banner" as being different versions of the same song, so too can we (in a way) think about the quantitative portions of each of these problems as being different versions of the same algebraic song identified through the algebraic expression (algebraic lyrics) $16x + 10 = 106$.

This hints at what French mathematician and scholar Henri Poincaré meant when he said, "mathematics is the art of giving the same name to different things" (where "name" is "equation" in this case).[36]

The algebraic ensemble is infinite in scope and undoubtedly includes many more possibilities, including real-world scenarios such as those in problems 4 and 5. Here, problems 4 and 5 correspond to cases where classroom word problems and real-world situations can overlap.

The basic algebra that we already know allows us to quickly solve all of these problems in an identical and familiar way:

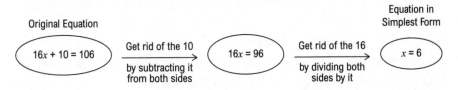

The solution then is 6, which interprets in each of the respective problems as follows:

1. The number 6.
2. The first piece of rope is 6 feet long, which cascades to 52 feet and 48 feet for lengths of the second and third pieces, respectively.
3. The rectangle's width is 6 meters, which cascades to 47 feet for its length.
4. 6 days.
5. 6 hours.

As Poincaré implies, the demonstration yields a solution that accurately applies to all of the various cases.

The symbolic tying together of a host of distinct objects and actions is not really a new concept, being one of the hallmarks of language itself. For example, the English word *chair* symbolically ties together millions of distinct objects of different sizes, weights, colors, and shapes. And the statement "the man walked down the road" ties together multiple trillions of distinct actions—any man (of any age, size, personality, etc.) walking down any road, for any length of time, on any given date (past or present) anywhere on the planet. And so on with other words and statements in English or in any other language. It is one of the things that gives language its great reach from a comparatively small glossary.

A huge difference, though, is that in the case of language most of us can naturally make the connection between a huge majority—if not all—of the objects we call chairs in spite of their diversity, and similarly with the different performances of "The Star-Spangled Banner." Connecting them through the lyrics is almost as natural as recognizing a familiar voice or face.

In the case of algebra, however, the connections are more cloaked and can rarely be made from sensory inspection only. Moreover, algebraic similarities exist on different levels—ranging from the exact algebraic connections (that simplify to the same equation) made between problems 1 through 5—as is the case in this section—to the slightly more general points of contact (that represent the same type of equation) made between the problems in Chapter 5.

Sometimes it is only through mathematics that we can see how intimately connected diverse problems are. This is one of the most beautiful things about the subject. The cloaking of connections continues and can become increasingly obscure and sophisticated as the subject advances, and math's ability to unearth even these similarities continues to inspire admiration and awe in those who bear witness to it.

A HEIGHTENED AWARENESS

The realization that many things which don't look or seem alike at all can still be tied together by common mathematical expressions, equations, and reasoning is a key ingredient in becoming more mathematically aware. This, combined with the knowledge that in many cases the only way to make such connections at all appears to be through mathematics, demonstrates that there is a lot going on out there in the world that we simply can't see without tuning into mathematics—"to fly where before we walked," as Bill Thurston proclaimed.[37]

Mathematics is a continuum, so its ability to connect many things together shows itself from the start, even in elementary arithmetic. For example, the simple operation 32 × 48 can answer questions such as

- the area of a rectangle with a width of 32 feet and a length of 48 feet,
- the number of chairs in an auditorium that has 48 rows each with 32 chairs,
- the amount of money earned working for 32 hours if the wage is $48 an hour,
- the distance a train travels for 32 hours if its average speed is 48 miles per hour.

But in each of these cases, the mathematical connection through multiplication is almost as straightforward to recognize as it is recognizing that the different performances of "The Star-Spangled Banner" all represent the same song.

It is in algebra where things start to regularly get intricate and complicated enough to the point where the mathematical connections between phenomena become less obvious and more hidden, requiring a different form of treatment.

This makes the subject uniquely positioned to give students a powerful taste of what the higher aspirations of mathematics and science are truly about—other things like reorganization, unification, systematization, experimentation, and discovery, in addition to numerical computation and symbolic manipulation. And because these higher aspirations are central to the continuity and operation of our modern technological and data-driven society, the opportunity for more people to experience this taste should not be easily dismissed or caricaturized.

Many years ago, Alfred North Whitehead spoke to the issues of requiring courses, such as the classics, in an increasingly crowded curriculum:

We must remember that the whole problem of intellectual education is controlled by lack of time. If Methuselah was not a well-educated man, it was his own fault or that of his teachers. But our task is to deal with five years of secondary-school life. [A course in the] classics can only be defended on the ground that within that period, and sharing that period with other subjects, it can produce a necessary enrichment of intellectual character more quickly than any alternative discipline directed to the same object.[38]

I believe that requiring algebra in the curriculum can also be defended on such elevated grounds. Its role in underwriting so much in our high-tech world, along with its ability to offer the proper mix of generality and concreteness, places it in a potentially commanding position of advancing powerful perspectives and insights across the disciplines—but only when utilized in its more optimal and expressive forms. The subject in education is not unlike a sophisticated airplane, which is capable of truly remarkable feats when operated at the proper speeds and altitudes, yet when flown improperly is also capable of stalling out, breaking apart, and suffering catastrophic failure.

The ability to see the impact of algebra—from taking only a course or two—on the wider world more often than not comes from a heightened awareness and comfort level with the subject. Though those who have had bad experiences with algebra may disagree, some of this heightened awareness and comfort may have already developed without them even noticing.

Consider the following: Last year there were 2,142,341 people who got married in the United States. How is this possible? Namely, how is it possible to have an odd number of people get married when there are two people involved in every marriage? It would seem that the number should always be even.

There are many ways to explain how this can happen—for instance, consider A, B, and C and imagine that A and B get married in January and then divorced in June. If B and C then get married in November while A remains single, we see that three people (an odd number) got married in a year.

Now, if you were able to follow this explanation without asking such questions as "How can three letters in the alphabet get married?" or "What in the world does an argument involving the first three letters of the alphabet have to do with 2,142,341 people getting married?", then you may have already taken something algebraic away with you from your math classes.

You most likely have taken away a certain comfort level, or fluency, with the notion of letters acting as placeholders for varying information like numbers or individuals. If you were able to understand how the marriages between A, B, and C might explain a number that represents millions of marriages, then you also likely have developed some familiarity with generalizing quantitative relationships. That is, this simple example exposes the flaw in the assumption that a single individual can't be involved in two different marriages in one year. From this, we realize that this circumstance could happen multiple times in a group numbering in the millions, allowing for the possibility of an odd number of people getting married.

To put it another way, we could also explain our resulting odd number if we simply said that someone could have gotten married, then got divorced, and then married somebody else in the same year while their first spouse remained unmarried. This reasoning involves unknowns and generalizations just as our algebraic-like rendering did—check to see how it relates to our initial framing involving the letters A, B, and C (and what is explicit and what is implied in the reasoning).

Heightened awareness exists in areas other than mathematics. For instance, say we have two people, one comfortable using the internet and the other not, who want to learn how to fix a flat tire on a car. The person who prefers traditional research only can certainly avail themselves of library resources, or try to find someone who can show them how to fix a flat tire. The other individual who is comfortable with the internet has these options plus the ability to do Google and YouTube searches, where they can find a variety of videos that discuss the process or show tires being changed on multiple types of cars. The latter person thus has more resources at their disposal and therefore a much better chance of successfully achieving their objective.

Consider a second example of two different people, one who has only basic internet literacy and the other who is more computer savvy. Say they both buy MP3 music players and want to transfer all of the stored music on their computers to these portable players, only to discover that the stored music is in a format—say RMJ, for Real Player software—that an MP3 player won't play. The computer-knowledgeable person who knows that files can be converted from one format to another—for instance, Word files and Excel files to PDFs, or Photoshop files to JPEGs—decides to search for a program that can convert RMJ files to MP3s. A Google search quickly reveals that such programs exist, and 20 minutes and $10 later, they are happily converting their files to MP3 files. This person's experience has provided a wider framework for them to situate a new problem and try to find a solution. The fact that this option even exists may never occur to the other person, and even if it does, they may lack the confidence and know-how to even download the files, much less carry out the actual conversion. They simply lack a broader contextual framework to resolve these types of situations.

These sorts of scenarios also occur when mathematical situations make planned and unplanned appearances in daily life, and people's lack of confidence in the subject intimidates them into unnecessary avoidance and inaction. In other circumstances, simple mathematical interactions may be present yet remain unrecognized, along with basic operations and interpretations that could clarify the issues. In either case, with a bit more confidence and overall algebraic awareness, the mathematically disinclined could realize that they have additional resources at their disposal to simplify a problem they're facing or learn more about it.

CONCLUSION

We have undertaken a very brief history of issues surrounding algebra's place in the secondary school curriculum, and in so doing, we considered one of its central tensions: the value of elementary algebra to a contemporary high school or adult student. Here, we have argued that even if a student's only exposure to the subject is a couple of classes in high school or college, algebra can be taught in such a way that creates a heightened awareness that will serve students in their future careers, whether in a STEM field or otherwise.

Next, we will explore this thesis in a bit more detail to see if we can really use any of the ideas from a course or two in elementary algebra in nontrivial ways to better understand and tune in to situations involving numerical variations that we may actually encounter—or hear about—in the course of our daily lives.

Motions in Life

Mathematics is the extension of common sense
by other means.

—Jordan Ellenberg, *How Not to Be Wrong:*
The Power of Mathematical Thinking

8

Algebra Uncloaked

We begin by revisiting the number of days and age problem. If you re-call from Chapter 2, the simplified formula that encoded the process for someone who has had a birthday in the current year is $100x + (2013 + z - y)$, where x is the number of days the person likes to eat out in a week, y corresponds to their year of birth, and z represents the number of years we are past 2013.

For simplicity's sake, we will assume that we are dealing with those who have had a birthday in the current year, and that the date is June 30 of that year. If z equals 8, this corresponds to a current year of 2021, so let's start there.

Substituting 8 for z simplifies the variable expression to $100x + (2021 - y)$. Here, z acts as a parameter that is constant for a given year, but changes from year to year.

EXPRESSING VARIATION WITH TABLES

We will now construct a table containing the values for the variable x, the variable y, and the variable expression $100x + (2021 - y)$. For the values $(x = 1, y = 1930)$, $(x = 2, y = 1952)$, and $(x = 6, y = 1965)$, we have the following:

Value of x	Value of y	Value of $100x + (2021 - y)$
1	1930	$100(1) + (2021 - 1930) = \mathbf{191}$
2	1952	$100(2) + (2021 - 1952) = \mathbf{269}$
6	1965	$100(6) + (2021 - 1965) = \mathbf{656}$

In the second example, for the personal three-digit number 269, 2 represents the number of days of the week the person likes to eat out and 69 represents the age of the person (born in 1952) whose birthday in 2021 is on or before June 30.

Recall that this particular problem has certain conditions that the variables must fulfill in order for the interpretation to work. For x, the value must be one of the whole numbers 1 through 7, and for y, the whole number values range from the year 1922 to the current year 2021—corresponding to individuals under the age of 100.[1] This means that the fully filled-in table will have 700 rows of data, each corresponding to the result of one of the possible number of days of the week and year of birth combinations.

For those who have yet to have a birthday by June 30, 2021, there will be a slightly different table in which the y values will range from the years 1921 to 2020.

Placing these values in a table demonstrates that we have another powerful way of expressing an ensemble of numbers. Not only are expressions able to capture and represent variable behavior, but tables can, too; and just as a specific word may be more appropriate than its synonym in a given context or sentence, so it is with these two different representations of numerically variable phenomena.

When we are recording quantitative data in an experiment, for instance, it can be helpful to first capture it in a table and see if it is possible to reverse-engineer an algebraic expression. In cases where that is hard to do exactly, a statistical approximation method known as regression may be employed to yield a formula that can estimate the data. In other situations, like the number of days and age problem, the procedure may be spelled out in such clarity that first capturing it as an algebraic formula may be the more straightforward way to proceed.

Greater algebraic awareness means becoming mindful of how algebra might structure or inform sets of data that we may encounter on the job or use in our everyday lives. Next time you encounter an IRS table, a retirement benefits table, or a motor vehicle depreciation table, consider the possibility that an algebraic expression may be responsible for generating those values behind the scenes.

SPREADSHEET AND VARIATION: FAST ALGEBRA

Let's look at the tabular analysis in the previous section as Microsoft Excel might view it. Excel—like most spreadsheet software—is an extremely efficient tool for synthesizing numerical information in tables and maneuvering it. The next table simply lists the raw data for someone who likes to eat out 1, 2, or 6 times a week in column A, and who was born in the years 1930, 1952, or 1965, respectively, in column B. Column C is blank for the moment.

A	B	C
1	1930	
2	1952	
6	1965	

Excel reads this table by cell addresses—for example, A3 means the cell location that is in column A and in the third row, which here contains the number 6. Similarly, B2 refers to column B and row 2, the cell that contains the value 1952. What we want to do is instruct Excel to populate column C with the values of the variable expression $100x + (2021 - y)$, where the entries in column A represent x values and the entries in column B represent y values.

Remember, computers cannot read as we do, meaning they can't just visually scan a row of values and place them into the appropriate part of the expression in column C. For them to be able to successfully locate and collate data, we have to input explicit instructions formatted in specific notation.

This can be done in Excel by first entering "=100*A1 + (2021 – B1)" in C1. [Note that the asterisk (*) represents multiplication in Excel.] This can be read as "100 times (entry in A1) + 2021 – (entry in B1)" or for our example, "100(1) + (2021 – 1930)," simplifying to 100 + 91 = 191, which corresponds to the value in our earlier table for $x = 1$ and $y = 1930$. Next, we apply this same formula for the entries in (A2, B2) and (A3, B3), placing the results respectively in C2 and C3—and so on up to the formula for (A700, B700) in C700.

Once the data for columns A and B have been entered, Excel enables us to apply the speed and vigor of its computational power to all 700 situations—almost instantaneously—to calculate the values in column C:

A	B	C
1	1930	100*A1 + (2021 – B1) = **191**
2	1952	100*A2 + (2021 – B2) = **269**
6	1965	100*A3 + (2021 – B3) = **656**
⋮	⋮	⋮

The expression "100*A1 + (2021 – B1) =" does not usually show up in the final result in C1 in Excel. Only the number 191 appears in C1, and likewise for C2, C3, and the other cells.

Thus, Excel creates a visual expression of the algebraic calculations that we discussed earlier in the number of days and age problem and the numerical ensemble generated by it. In other words, at its core, some of Excel's processes are really a fast, pictorial type of algebra!

COMPUTING GRADES

Let's look at another numerically variable situation with algebraic properties masked by familiarity.

Consider the case of an instructor calculating individual grades for a math class of 30 students. Throughout the course, students typically are expected to complete homework assignments, several tests during the term, and a final exam. Each of these three components contributes to the overall grade for the course, but given that an hour-long, proctored test usually has more stringent demands and expectations on students than homework assignments do, many math instructors place a higher value on test scores.

In other words, to figure out a student's final grade, instructors typically use an uneven mixture of these three categories of assessment and assign each one a different contribution strength—or weight—based on its relative importance to the overall grade. The particular weight of each category will, of course, depend on the preferences of each individual instructor. Let's assume that this professor decides that the average score on all of the homework assignments will contribute 25% to a student's overall final grade, while the average score on all of the tests will contribute 55% and the score on the final exam will be the last 20%.

This situation has a lot in common with the problems we have already encountered. First of all, it is variable, because each student will have average scores for each of these categories that change depending on the student. So, instead of students having varying selections for the number of days of the week they like to eat out, their year of birth, and the years past 2013 in the problem, each student now is effectively making selections for three new variables: their homework average score, their average test score, and their final exam score.

Now the three scores don't seem like selections in the same sense as selecting the number of days of the week you like to eat out does, because the student has long-range interaction with each of these scores and doesn't simply "pick" them at will—rather the student "earns" these numerical values through their very effort and comprehension of the material. Nevertheless, these scores are dependent upon variable student performance—and look as such to the instructor—meaning that now the students, themselves, have become part of the variation in the problem. These possible scores range from 0 to 100, rounded to the nearest whole number.

Placing all of the average scores for six such students in a table yields the following:

Name	Homework Average	Test Average	Final Exam Score	Numerical Course Grade
Hypatia	96	95	99	
Al-Samawal	90	92	94	
Dardi	88	90	93	
Ramus	96	88	54	
Franklin	30	50	95	
Darwin	78	68	72	

This table represents part of the numerical symphony of scores for the entire class. As the scores vary from student to student, we can represent each category by tagging them with a different letter. Let's say that x = homework average, y = test average, and z = final exam score.

The full numerical course grade for each student in the class then depends on how these average scores are weighted. We can use the algebraic expression $0.25x + 0.55y + 0.20z$ to analyze how mixing together these three types of assessments at the prescribed strengths produces this grade. From now on, we will call the numerical course grade for a student their course average. Note that in this formula we have substituted the decimal equivalents for the percentages: 0.25 for 25% (homework), 0.55 for 55% (tests), and 0.20 for 20% (final). We will switch back and forth between these equivalents depending on the situation.

The following table includes the computed course averages for the six students:

Name	x	y	z	Course Average = $0.25x + 0.55y + 0.20z$
Hypatia	96	95	99	96
Al-Samawal	90	92	94	92
Dardi	88	90	93	90
Ramus	96	88	54	83
Franklin	30	50	95	54
Darwin	78	68	72	71

Using these course averages and the 10-point scale for letter grades will yield grades of A for Hypatia, Al-Samawal, and Dardi, a B for Ramus, and a C for Darwin, while Franklin would receive an F.

Range	Letter Grade
90–100	A
80–89	B
70–79	C
60–69	D
0–59	F

10-point scale

Of course, the standard way this data is represented rarely refers back to the x, y, and z variables of schoolroom elementary algebra. In Excel, this table plays out as follows:

A	B	C	D	E
Name	Homework Average	Test Average	Final Exam Score	Course Average
Hypatia	96	95	99	0.25*B2 + 0.55*C2 + 0.20*D2
Al-Samawal	90	92	94	0.25*B3 + 0.55*C3 + 0.20*D3
Dardi	88	90	93	0.25*B4 + 0.55*C4 + 0.20*D4
Ramus	96	88	54	0.25*B5 + 0.55*C5 + 0.20*D5
Franklin	30	50	95	0.25*B6 + 0.55*C6 + 0.20*D6
Darwin	78	68	72	0.25*B7 + 0.55*C7 + 0.20*D7
⋮	⋮	⋮	⋮	⋮

Pre-calculation

A	B	C	D	E
Name	Homework Average	Test Average	Final Exam Score	Course Average
Hypatia	96	95	99	96
Al-Samawal	90	92	94	92
Dardi	88	90	93	90

(*table continued on next page*)

(table continued from previous page)

A	B	C	D	E
Ramus	96	88	54	83
Franklin	30	50	95	54
Darwin	78	68	72	71
⋮	⋮	⋮	⋮	⋮

Post-calculation

This setup is general and can be applied to class after class over the course of an instructor's career, ultimately leading to a numerical ensemble of individual quantitative variation numbering in the thousands of students.

BIG ALGEBRA RETURNS

This is just one scenario for one instructor. Imagine a different instructor who—perhaps believing it unfair to students with test anxiety to favor tests over homework, where they can demonstrate their comprehension of the material under less pressure—decides to assign different weights to each category. What if this instructor's grading policies are that a student's average score on homework will count for 60% of their final grade, their average score on tests will be 30%, and the final exam will count for 10%? Assuming that x, y, and z remain the same, the formula to find a student's final course average would change to $0.60x + 0.30y + 0.10z$.

The table for final course averages now reads as follows:

Name	x	y	z	Course Average = $0.60x + 0.30y + 0.10z$
Hypatia	96	95	99	96
Al-Samawal	90	92	94	91
Dardi	88	90	93	89
Ramus	96	88	54	89
Franklin	30	50	95	43
Darwin	78	68	72	74
⋮	⋮	⋮	⋮	⋮

The Excel analysis and formatting from the first case would apply here as well, with the only difference being the formula in Column E, which would be 0.60*B2 + 0.30*C2 + 0.10*D2 for the entry in the second row and so on:

A	**B**	**C**	**D**	**E**
Name	Homework Average	Test Average	Final Exam Score	Course Average
Hypatia	96	95	99	96
Al-Samawal	90	92	94	91
Dardi	88	90	93	89
Ramus	96	88	54	89
Franklin	30	50	95	43
Darwin	78	68	72	74
⋮	⋮	⋮	⋮	⋮

These results will now yield grades of A for Hypatia and Al-Samawal, B for Ramus and Dardi, and C for Darwin, while Franklin would still receive a grade of F.

Yet a third instructor who believes that the final exam is the best assessment of a student's mastery of all of the material in the course might decide to give it much more weight. Let's say that this instructor decides that a student's average homework score counts for 15%, their average test score counts for 25%, and their final exam counts for 60%. The table for final course averages now reads as follows:

Name	x	y	z	**Course Average =** $0.15x + 0.25y + 0.60z$
Hypatia	96	95	99	98
Al-Samawal	90	92	94	93
Dardi	88	90	93	92
Ramus	96	88	54	69
Franklin	30	50	95	74
Darwin	78	68	72	72
⋮	⋮	⋮	⋮	⋮

This results in A's for Hypatia, Al-Samawal, and Dardi, C's for Darwin and Franklin, and a D for Ramus—certainly a vastly better outcome for Franklin because he did so well on the final.

These different scenarios are reminiscent of the different scenarios for a business in Chapter 4, where instead of the selling price per unit, cost per unit, and overhead fixing a given scenario, we have the percentage contributions assigned to homework, tests, and the final now fixing a specific scenario.

This suggests that parameters may again be useful to describe not just one specific scenario but an entire class of them. We can use letters earlier in the alphabet to represent each of the percentage contributions as follows: $a \equiv$ percentage of contribution from homework to course average, $b \equiv$ percentage of contribution from tests to course average, and $c \equiv$ percentage of contribution from the final exam to course average. If we continue to use x, y, and z to stand for a student's scores in each respective category, the general expression for their course average for all possible scenarios is $ax + by + cz$.

The following table summarizes all three instructors' grading policies:

Scenario	Weights/Contributions	Parameter Values	Course Average $= ax + by + cz$
1	25% from homework 55% from tests 20% from final exam	$a = 25\%$ $b = 55\%$ $c = 20\%$	$0.25x + 0.55y + 0.20z$
2	60% from homework 30% from tests 10% from final exam	$a = 60\%$ $b = 30\%$ $c = 10\%$	$0.60x + 0.30y + 0.10z$
3	15% from homework 25% from tests 60% from final exam	$a = 15\%$ $b = 25\%$ $c = 60\%$	$0.15x + 0.25y + 0.60z$

Scenario table

One might now naturally ask, why go through the mathematical gymnastics to create a formula like this? Don't most teachers know how to calculate their students' final grades for their class without considering parameters at all?

Establishing parameters enables any instructor—who uses these three assessment categories and wants to calculate course averages—to tune the parameter values to match their grading criteria, which will give them the correct algebraic expression for the specific circumstances of their class. All that's left is to input each student's averages for x, y, and z. This aligns well with Excel and other software where a macro or app can be created that will empower an instructor to swiftly calculate all of the course averages for their class, seemingly all at once.

Using parameters also expands our conceptual understanding of the algebraic processes at play in computing grades. In the next sections, we will see that these processes are really representatives of an entire genre of algebraic songs with useful interpretations and renditions in many other areas.

QUANTITATIVE COCKTAILS

Our three instructors, in calculating course student averages by weighting different categories of assessment, have actually been doing something procedurally similar to a bartender mixing cocktails by combining various ingredients in different strengths. In other words, we can think of the procedures for calculating final course averages as the making of various types of "quantitative cocktails" out of different mixtures of homework averages, test averages, and final exam scores. The next table compares the recipe for our final course average "cocktail" to a piña colada cocktail, highlighting the three most important ingredients in each.

Thus, for Scenario 1 we would have the following mixture:

	Piña Colada Cocktail	Final Course Average Cocktail
Ingredient 1 (25%)	White rum	Homework average
Ingredient 2 (55%)	Pineapple juice	Test average
Ingredient 3 (20%)	Coconut cream	Final exam score

Using parameters, we could summarize all of the scenarios as follows:

	Piña Colada Cocktail	Final Course Average Cocktail
Ingredient 1 ($a\%$)	White rum	Homework average
Ingredient 2 ($b\%$)	Pineapple juice	Test average
Ingredient 3 ($c\%$)	Coconut cream	Final exam score

Quantitative cocktails are, however, a bit different in the details from their alcoholic counterparts. Let's first look at Scenario 1 where the overlap is more pronounced. Consider a quart container in which we wish to make a piña colada with the stated percentage mixture of ingredients, and suppose that we have three smaller containers each with the capacity of 25%, 55%, and 20% of a quart. We fill each of these to the brim with white rum, pineapple juice, and coconut cream, respectively. If we then pour the contents of each of these three smaller containers into the larger container and mix them, we create a full quart-size piña colada. That full piña colada correlates to a student with a perfect score in all three assignment categories—meaning that they receive the full 100 percentage points for the course average.

But what happens if a student doesn't have a perfect average on each category of assignment, such as with Ramus? To see how the metaphor looks for someone with less than perfect scores, we continue with the perfect scores situation.

Let's think of this situation as one where we have an initially empty 1-quart container representing the final course average and three smaller containers—the 25%, 55%, and 20% quart containers—which correspond, respectively, to homework average, test average, and the final. Now imagine each of these smaller containers as being filled with a fluid of the same type, in contrast with the piña colada case, where the constituents are different in each container. Let's take this fluid here to be water.

So, for the student with perfect scores, the three smaller containers are filled to the brim with water, meaning that when they are each emptied into the quart container, this larger container is filled 100% to the brim.

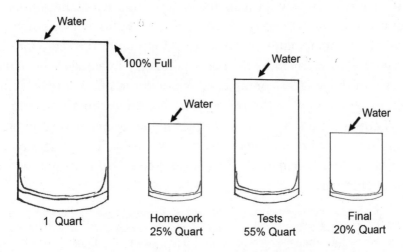

[Artwork provided courtesy of William Hatch]

For the student with less than perfect scores, each of the smaller containers are not completely filled with water, and thus when they are combined in the larger container, the amount of liquid isn't a full quart. In Ramus' case, the homework container is 96% full, the test container is 88% full, and the container for the final exam score is filled up 54% of the way. Pouring these quantities of liquid together into the larger quart container to mix Ramus' quantitative cocktail, we find that it's filled up to 83% of its total capacity. This, of course, corresponds to Ramus' course average, or a B grade.

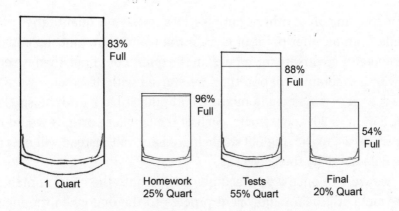

[Artwork provided courtesy of William Hatch]

If it were possible to have digital readouts on each of the containers that showed the percentage of liquid in each container—rounded off to whole number values—then we could represent the score for each category of assessment by the corresponding percentage level of fluid in each of the smaller containers. We could then find out the course average by combining the water from the three containers into the quart container and reading its display in lieu of calculating on paper or in Excel. In short, mixing the fluids this way could serve as a type of special-purpose calculator.[2] The table for Scenario 1 is given here for a quart container:

A	B	C	D	F	E
Name	Homework Average Container	Test Average Container	Final Exam Score Container	Quart Container	Course Average
Hypatia	96% full	95% full	99% full	96% full	96
Al-Samawal	90% full	92% full	94% full	92% full	92
Dardi	88% full	90% full	93% full	90% full	90
Ramus	96% full	88% full	54% full	83% full	83
Franklin	30% full	50% full	95% full	54% full	54
Darwin	78% full	68% full	72% full	71% full	71
⋮	⋮	⋮	⋮	⋮	⋮

INTERPRETATIONS

Before moving on to other examples, let's revisit our quantitative piña colada from another point of view. If the white rum, pineapple juice, and coconut cream are mixed well in the quart container, then we end up with a different situation than we started with. If we now were to take the quart piña colada mixture and pour it back into the smaller 25%, 55%, and 20% containers, each of the smaller portions would no longer consist of the original single ingredient, but instead will now be piña coladas all by themselves.

If we were to once again recombine them into the quart container, this would still be an equivalent process to the one when the ingredients were separate. This illustrates the essence of what the initial blending of the ingredients tells us. That is, we can think of the blending of the three separate ingredients as an operation on these ingredients that produces a new combined object—the cocktail drink itself.

Applying this perspective to Ramus' fluid levels, we know that mixing together the 96%, 88%, and 54% full smaller containers yields a quart container that is 83% full. We could now reapportion this fluid into each of the smaller containers in such a way that each is an individual copy of the larger container in that it is 83% full, too. This is one of the things that this blended average tells us—83% perfectly blends all of Ramus' scores over the semester. That is, if he received a score of 83 on every single assignment that he did that semester—all of the homework assignments, all of the tests, and the final—then his average would match the average he actually received with all of the varying individual scores and differently weighted assessments.

Though receiving an identical score on every assignment didn't, of course, happen for Ramus—unlike a well-mixed cocktail where perfect blending can occur—it still can be very useful to think of the collective average in this way. We will leverage the ideas in the last few sections where appropriate.

CLASS SIZES AS PARAMETERS

Three professors give the same common final exam to their respective Math 116 classes: one with 25 students (Section 001), the second with 55 students (Section 002), and the third with 20 students (Section 003). Before leaving for summer break, the professors each record their class averages on the exam as follows:

Section	Final Exam Average
001	96
002	88
003	54

Later during the summer, the department chair wants to obtain an overall average on the final exam for the students in all three sections to report to college administration. Unfortunately, she doesn't have the individual test scores for each student, meaning that she can't find the average by simply adding up all 100 individual scores and then dividing by 100. Now that the three professors are on vacation, she'd rather not bother them to get the scores, so she must figure out another way.

One option would be to simply calculate the average of the three class averages. Doing this yields

$$\frac{96+88+54}{3} = 79.333 \ldots,$$

which rounds off to a 79 average.

The problem with this approach is that there were more than twice as many students—and therefore twice as many test scores—in Section 002 than in either of the other two sections, so its average should carry more impact or weight, and the current calculation doesn't give it that. How can we factor in this greater impact?

One thing the department chair does have is easy access to the student enrollment numbers for each class. She adds this information to the table:

Section	Exam Average	Number of Students
001	96	25
002	88	55
003	54	20
Total Number of Students		100

What the department chair wants to do is to assign a value to each class based on the relative size of its student population, where a class with more students is given greater weight. This turns out to be another type of quantitative cocktail where, instead of the ingredients being three types of assessments with different contribution values, they are three final exam averages for three classes with different enrollment values.

	Piña Colada Cocktail	Final Course Average Cocktail	Final Exam Average Cocktail
Ingredient 1	White rum	Homework average	Section 001 average
Ingredient 2	Pineapple juice	Test average	Section 002 average
Ingredient 3	Coconut cream	Final exam score	Section 003 average

This time, however, instead of the department chair having the power to determine the concentration of each ingredient, there is a straightforward way to decide what the contribution for each class should be. She can simply calculate the proportion of students in a given class out of the total number of students who took the course:

$$\% \text{ contribution for section} = \frac{\text{number of students in class section}}{\text{total number of students in all three sections}}$$

$$= \frac{\text{number of students in class section}}{100 \text{ students}}.$$

Computing this for each class gives their weight:

	Final Exam Average Cocktail	Number of Students	Section Contribution to Overall Exam Average	Final Exam Average
Ingredient 1	Section 001	25	0.25 or 25%	96
Ingredient 2	Section 002	55	0.55 or 55%	88
Ingredient 3	Section 003	20	0.20 or 20%	54

The percentage concentration of each ingredient corresponds to the parameter values, meaning that $a = 0.25$, $b = 0.55$, and $c = 0.20$ as in Scenario 1 from the previous sections of this chapter. The regular variables are the averages for the final exam in each class, corresponding to $x =$ final exam average in Section 001, $y =$ final exam average in Section 002, and $z =$ final exam average in Section 003.

Thus, the expression $ax + by + cz$ becomes $0.25x + 0.55y + 0.20z$, and after substituting the final exam averages with $x = 96$, $y = 88$, and $z = 54$, the department chair can calculate that the final exam average for all 100 students is 0.25(96) + 0.55(88) + 0.20(54) = 83 (rounded off).

Note that this value differs from the final exam average of 79 that we obtained when each class contributed equally to the average. It's also worth noting that if the department chair had been able to add all one hundred individual scores and then divide by 100, she would have arrived at the same value of 83, or close to it.[3] We can think of this 83 as being the score that all of the different individual scores of the 100 students blend to when we mix them together—as if each and every one of them got the identical score of 83 on the final.

The department chair has performed a powerful mathematical maneuver in calculating the overall final exam average without actually seeing a single individual student exam score. It doesn't mean, however, that the individual student exams and scores were never seen in the process—obviously, the three professors saw the individual exams in each of their respective classes and graded them accordingly. What the technique employed here allowed the department chair to do, was to piggyback off of the efforts of her colleagues in a nontrivial way to gain new and useful information. That is, she acquired it by using an algebraic formula to process the three class averages differently, completely bypassing the need to do a direct computation involving 100 numbers.

This is conceptually similar to how the formula for the number of days and age problem enabled us to generate those special three-digit numbers in place of following each individual step in the original procedure presented in Chapter 1.

Observe that the department chair's situation is mathematically identical to the calculation for Ramus' course average in Scenario 1. These two Excel tables demonstrate this:

A	B	C	D	F
Name	Homework Average	Test Average	Final Exam Score	Course Average
Ramus	96	88	54	0.25*B2 + 0.55*C2 + 0.20*D2 = **83**
Course	Section 001 Average	Section 002 Average	Section 003 Average	Final Exam Average for All Three Sections
Math116	96	88	54	0.25*B5 + 0.55*C5 + 0.20*D5 = **83**

We can see how these two particular versions of the same type of mathematical idea—or song—play out. The first produces a single number that represents an individual student's performance derived from their various scores across the three categories of homework, tests, and the final exam. The other generates a single number that represents the results of many individual students on one final exam across three class sections.

Before moving on, it is worth pointing out again that these two renditions of this idea differ in the way that the contribution weights are assigned. In the case of the individual course average for a student, the instructor had quite a bit of freedom to weight the value of each assessment category as they saw fit, influenced by factors such as school policies and their own educational philosophy. In the second case of the collective final exam average for three classes, the contributions were already predetermined for the department chair on the basis of a particular section's enrollment out of 100 students. The existence of such inherent parameter values often occurs with this type of problem.

In the end, however, mathematics doesn't ultimately care how the parameters were obtained. Once they have been established, the math involved in the calculation takes care of itself.

A THIRD VERSION: GRADE POINT AVERAGE (GPA)

Imagine that a student has just received their report card for spring semester, earning a semester GPA of 2.75:

Course	Grade	Credits
Chemistry	A	4
Calculus	C	4
Economics	B	4
Composition	C	4
Spring Semester GPA: 2.75		

How would we perform a calculation by hand to verify that the GPA is correct?

One formula for calculating GPA reads as follows:

$$\frac{\begin{array}{c}(\textbf{A credits})\cdot(\text{A points})+(\textbf{B credits})\cdot(\text{B points})+(\textbf{C credits})\cdot(\text{C points})\\+(\textbf{D credits})\cdot(\text{D points})+(\textbf{F credits})\cdot(\text{F points})\end{array}}{\text{total credits}},$$

where

$$A = 4 \text{ grade points,}$$
$$B = 3 \text{ grade points,}$$
$$C = 2 \text{ grade points,}$$
$$D = 1 \text{ grade point,}$$
$$F = 0 \text{ grade points}$$

GPA Formula (1)

From the student's report card, we see that they have taken a total of 16 credits for the spring semester, 4 of which were at an A grade, 4 at a B grade, 8 at a C grade, 0 at a D grade, and 0 at an F grade. Plugging this information into the GPA formula, along with the point values for each grade, yields the following for the student's spring semester GPA:

$$\frac{\begin{array}{c}(\textbf{4 credits})\cdot(4\text{ points})+(\textbf{4 credits})\cdot(3\text{ points})+(\textbf{8 credits})\cdot\\(2\text{ points})+(\textbf{0 credits})\cdot(1\text{ points})+(\textbf{0 credits})\cdot(0\text{ points})\end{array}}{16\text{ credits}}$$

$$=\frac{(16+12+16)\text{ grade points}}{16\text{ credits}}=\frac{44\text{ grade points}}{16\text{ credits}}=\frac{2.75\text{ grade points}}{1\text{ credits}}$$

$$=2.75\text{ grade point average.}$$

This 2.75 GPA matches the report card.

What this calculation does is help us find the contribution value of each grade by adding all of the credits under the umbrella of that grade and then multiplying its numerical point value and credits. The varying contribution values for each letter grade are then blended together—for this student, that means 16 grade points for the A in Chemistry, 12 grade points for the B in Economics, and 8 grade points each for the C's in Calculus and Composition—to yield the total grade points for the semester. These points are then divided by the total credits taken to give the GPA, illustrating that a GPA is just another type of quantitative cocktail with different numerical ingredients blended together in varying concentrations.

We can interpret this GPA by observing that the 44 total grade points for the semester out of 16 semester credits blend to 2.75 grade points per credit, out of 4 possible grade points per credit. Thus, if we multiply 16 credits times the 2.75 grade points per credit, we get back the 44 total grade points.

The uneven distribution of 44 grade points (among the one A, one B, and two C grades) gets blended to this 2.75 grade points per credit and by extension to the 2.75 grade points per course. We can think of the student as having made slightly less than a B grade in each of the four classes.

In its present form, however, the formula may seem a bit mysterious and isn't an exact corollary to the earlier examples of quantitative cocktails that we've worked through. In both of those examples, we had a total of something—such as 100% of a grade or 100 students—spread out over categories like "types of assessment" or "class sections." The distribution of this spread defined the scenario, which we then captured by tuning the parameters to the appropriate values. In both cases, the values of the parameters (a, b, c) and the variables (x, y, z) were each represented as percentages.

The grade point values used to calculate a GPA are generally not used that way. If we viewed these values as percentages, we would have that an A would be 100% on a 4-point scale, a B would be 75%, a C 50%, a D 25%, and an F 0%. Thus, a 2.75 GPA on the percentage scale would be 68.75%, corresponding to 2.75 out of 4 as a percent. Note that these

percentages are different from the percentages that we generally asso-
ciate with each letter grade on the grading scale in an individual class.
Because our goal here is to use algebra as an aid in understanding, we
will stay with the standard way of representing a GPA for a 4-point
scale. So how, then, do we interpret this formula as a quantitative cocktail in the sense of the previous two examples?

Let's first see if we can maneuver the formula to calculate GPA a bit,
in order to better understand how it works. We need to use a property
of fractions to do this, so let's first review how to add two or more fractions with the same denominator:

Example 1:
$$\frac{3}{7} + \frac{2}{7} = \frac{5}{7}.$$

Example 2:
$$\frac{1}{8} + \frac{3}{8} = \frac{4}{8} \text{ or } \frac{1}{2}.$$

Example 3:
$$\frac{5}{24} + \frac{1}{24} + \frac{7}{24} + \frac{4}{24} + \frac{2}{24} = \frac{19}{24}.$$

These are several instances of an ensemble of numerical expressions,
so we can use algebra to represent them all in a single form as

$$\frac{x}{w} + \frac{y}{w} + \frac{z}{w} + \frac{u}{w} + \frac{v}{w} = \frac{x+y+z+u+v}{w}.$$

For example 1, we have $x = 3$, $y = 2$, $z = 0$, $u = 0$, $v = 0$, and $w = 7$.
For example 2, we have $x = 1$, $y = 3$, $z = 0$, $u = 0$, $v = 0$, and $w = 8$.
For example 3, we have $x = 5$, $y = 1$, $z = 7$, $u = 4$, $v = 2$, and $w = 24$.

This expression states the well-known arithmetic principle that
when you add fractions with a common denominator, you simply add
the numerators and retain the common denominator.

We have chosen to give the addition of five fractions here, as the general example, because this corresponds to the five fractions that we will

get from GPA Formula (1). Recall from Chapter 3 that this algebraic formula is an identity rather than an equation to be solved.

An effective way to think about the equals sign in an identity is like a simple doorway. You can come into a room one way through a door, and you may leave the room through the same door going the opposite way. Often when an identity is presented to us, the tendency is to read it from left to right, the way we read in English. However, unlike English, we can also read an identity from right to left as both sides are equivalent. To remind ourselves of this, we can simply rewrite it with the left-hand side becoming the right-hand side and vice versa:

$$\frac{x+y+z+u+v}{w} = \frac{x}{w} + \frac{y}{w} + \frac{z}{w} + \frac{u}{w} + \frac{v}{w}.$$

This is the principle we need to be able to maneuver GPA Formula (1) into a more familiar form. Before applying this identity, let's first simplify the formula by plugging in the grade point values for each letter grade:

$$\frac{\begin{array}{c}(\textbf{A credits})\cdot(4)+(\textbf{B credits})\cdot(3)+(\textbf{C credits})\cdot(2)\\+(\textbf{D credits})\cdot(1)+(\textbf{F credits})\cdot(0)\end{array}}{\text{total credits}}.$$

Now, applying the fraction identity [where the common denominator w = total credits and the numerators x = (A credits) · (4), y = (B credits) · (3), z = (C credits) · (2), u = (D credits) · (1), and v = (F credits) · (0)] yields the following:

$$\frac{(\textbf{A credits})\cdot(4)}{\text{total credits}} + \frac{(\textbf{B credits})\cdot(3)}{\text{total credits}} + \frac{(\textbf{C credits})\cdot(2)}{\text{total credits}}$$

$$+ \frac{(\textbf{D credits})\cdot(1)}{\text{total credits}} + \frac{(\textbf{F credits})\cdot(0)}{\text{total credits}}$$

GPA Formula (2)

Looking at the first term, we can rewrite $\dfrac{(\textbf{A credits}) \cdot (4)}{\text{total credits}}$ as $\dfrac{\textbf{A credits}}{\text{total credits}} \cdot 4$. For example, $\dfrac{8 \cdot 3}{4}$ can be written as $\dfrac{8}{4} \cdot 3$, where each simplifies respectively to $\dfrac{24}{4}$ and $2 \cdot 3$, which in turn both equal 6.

The fraction $\dfrac{\textbf{A credits}}{\text{total credits}}$ is simply the percentage of credits taken that were A credits out of the total number of credits in the semester. For the spring report card, this becomes $\dfrac{4}{16}$, which is 0.25 or 25%. We can think of the four other fractions in this same fashion as the percentage of credits that were B credits and so on. Substituting this information into GPA Formula (2) and moving the grade point numbers in front of each term gives the following:

4(% of A credits) + 3(% of B credits) + 2(% of C credits) +
1(% of D credits) + 0(% of F credits)

GPA Formula (3)

This time, the fixed elements of the expression are the grade point values and the variable elements are the percentages of credits received under the umbrella of each letter grade. So, we can let these fixed values define the scenario in this case—it would be a different scenario if we gave plus and minus values for grades such as A+ or A−.

Returning to the parameter form of the previous quantitative cocktails, this time with five categories instead of three, we obtain the formula $ax + by + cz + du + ev$. Substituting in the parameter values for the worth of each letter grade, the expression simplifies to $4x + 3y + 2z + 1u + 0v$.

It is important to note that in this case using constant grade point values as parameters means that, for this particular interpretation, they do not add up to 100%; rather, it is the variables, as percentages of credits taken for a certain grade, that add up to 100%. Thus, in this sense, the roles of parameters and variables are reversed for GPA Formula (3). The variables are now as follows:

x = Percentage of credits taken where an A grade was received

y = Percentage of credits taken where a B grade was received

z = Percentage of credits taken where a C grade was received

u = Percentage of credits taken where a D grade was received

v = Percentage of credits taken where an F grade was received

Let's now apply GPA Formula (3) to another report card, this time to a case where the student is taking more courses and the courses are worth different amounts of credits:

Course	Grade	Credits
Geology	A	4
Calculus III	F	4
Anatomy/Physiology	B	5
Computer Science	C	4
Political Science	A	3
Total Credits		**20**
Fall Semester GPA: ??		

Fall semester report card

Variables	Total Credits	Percentage (Out of 20 Credits)
x = total A grade credits	7	35%
y = total B grade credits	5	25%
z = total C grade credits	4	20%
u = total D grade credits	0	0%
v = total F grade credits	4	20%

Fall semester credit totals

Placing this information into the expression $4x + 3y + 2z + 1u + 0v$ yields

fall semester GPA = $4(0.35) + 3(0.25) + 2(0.20) + 1(0) + 0(0.20) = 2.55$.

You can check the calculation of the GPA this way by using the original formula, GPA Formula (1), to verify that they match.

Formatting the GPA formula into an explicit parameter form has its advantages, as it enables us to structure this calculation in a familiar conceptual framework and leverage our understanding of the first two examples. If we have a firm grasp on how earlier quantitative cocktails were mixed, we can cross-apply that understanding to this circumstance.

This is not unlike how we use percentages to transport understanding in arithmetic. Percentages provide a common framework to understand the relationship between parts and a whole. When we are confronted with an unfamiliar numerical relationship—such as 46,659 people voting in an election out of 155,530 eligible voters—percentages set up a ratio that enables us to understand that relationship in the familiar context of parts out of 100. In this case, that ratio would be $\frac{46,659}{155,530}$, which equals 0.30. We can interpret this as 30%, which means that the ratio of the number of people who voted to the total number of people eligible to vote is the same as if 100 people were eligible and only 30 of them voted.

So, if the parameter viewpoint in GPA Formula (3) gives us a similarly standard frame of reference, why use the original formula at all?

Though the parameter viewpoint may offer some conceptual advantages, the original GPA Formula (1) has certain computational advantages that become apparent when the ratios do not yield a simple decimal. For example, if the Anatomy/Physiology course were only worth 4 credits instead of 5, the total number of credits in the student's fall semester would be 19 instead of 20. The percentage of A credits to total credits would then be 7 out of 19, or 0.3684210526.... To use this number in the parameter formula, we would most likely want to round it off to something like 0.3684 or 36.84%, which would result in a loss of information for the calculation. Because the prime number 19 is not a very decimal-friendly denominator, the percentages for the other grades would have the same issue, amplifying the scale of information loss in the problem and ultimately leading to a less accurate result for the resulting GPA.

GPA Formula (1) avoids these issues by dealing with whole numbers throughout the majority of the calculation, saving rounding off until the final step after the division is performed. This will give it a more accurate answer in cases such as the 19 credits variant, where we could end up rounding numbers as many as five times if we were to use GPA Formula (3). So, it's best to use the generally more straightforward GPA Formula (1) once we understand the concepts that underpin the calculation of a GPA.

It is also worth pointing out that using the fraction maneuver on the GPA calculation here would have been difficult to perform or understand for someone who only knew how to compute numerical fractions with a calculator. In order to successfully understand and work with fractions when variable quantities are involved, we need a deeper understanding of fractions and what the operations mean, which is just one illustration of how arithmetic is such an important conceptual foundation for algebra.

A FOURTH VERSION: EFFECTIVE PERCENTAGE RATE OF RETURN ON INVESTMENTS

An investor is considering whether or not to invest their money in three stocks or make a less volatile investment that guarantees them a 4% return over the next year. They know that stocks come with the potential for higher gains, but they come with more risks too, some of which could lead to a lesser return or even a loss. The investor has a computer application that allows them to forecast different scenarios to see how much their money will increase from spreading it out over the three stocks, each with a possibly different percentage return. Though pleased with the program, it seems like a bit of a black box to them, and they want to know more about how the returns on investment are calculated. Over the course of their research, they find the following formula for calculating the effective percentage rate of return:

(% of money invested in Stock A) · (% return from Stock A)
+ (% of money invested in Stock B) · (% return from Stock B)
+ (% of money invested in Stock C) · (% return from Stock C).

Let's say that the investor places 25% of their money in the stock for Company A, 55% in Company B's stock, and the final 20% of their money in Company C's stock—with expected percentage returns of 3.0%, 5.0%, and 9.5%, respectively. Then, the effective percentage increase of their investment can be calculated as

$$(0.25) \cdot (3.0\%) + (0.55) \cdot (5.0\%) + (0.20) \cdot (9.5\%) = 5.4\%.$$

This gives an effective percentage rate of return of 5.4% on all of the money in this scenario, meaning that if they invested $50,000 for the year, their money would have increased by 50000(0.054) = $2700 in the stocks versus 50000(0.04) = $2000 with the safer investment.

The formula in this format helps the investor understand the situation a bit more than simply plugging numbers into the application, but they still don't have a firm handle on it. The calculation looks both familiar and strange to them at the same time.

Firstly, the investor is taking a certain amount of money and distributing it over three stocks. In the case of a $50,000 investment, that distribution would be $12,500 (25%) in Stock A, $27,500 (55%) in Stock B, and $10,000 (20%) in Stock C. The calculation shows them the overall projected increase in their finances, blending the results of their three individual investments expected to increase by 3.0%, 5.0%, and 9.5%, respectively. This suggests that once again we are dealing with a quantitative cocktail. The following table compares this to the other renditions of quantitative cocktails in the chapter:

	Final Course Average Cocktail	Final Exam Average Cocktail	GPA Cocktail	Effective Return Cocktail
Ingredient 1	Homework Average	Section 001 Average	% of credits receiving an A	% return of Stock A
Ingredient 2	Test Average	Section 002 Average	% of credits receiving a B	% return of Stock B
Ingredient 3	Final Exam Average	Section 003 Average	% of credits receiving a C	% return of Stock C
Ingredient 4			% of credits receiving a D	
Ingredient 5			% of credits receiving an F	

Thinking back to earlier cocktails, let's take a look at how this one is mixed and establish our parameters and variables. The levers that we can adjust to fix the scenario are the percentages of money invested in Stock A, Stock B, and Stock C. The values that change, then, are the increases and decreases in each stock's performance. Thus, we will consider the former as parameters and the latter as variables. For the expression $ax + by + cz$, this results in the following interpretation:

a = (percent of total amount of money invested in Stock A),
b = (percent of total amount of money invested in Stock B),
c = (percent of total amount of money invested in Stock C),
x = (percent return from Stock A),
y = (percent return from Stock B),
z = (percent return from Stock C).

For the investment scenario here, we already know that $a = 0.25$, $b = 0.55$, and $c = 0.20$, and so the formula becomes $0.25x + 0.55y + 0.20z$—which should look familiar because this, too, is operationally identical to Scenario 1 for the course average cocktail.

Let's now consider six investors each investing in a portfolio of three stocks, each with a different mixture of stocks than the other five. For example, Al-Samawal invests in three different stocks than Ramus, but they both allocate their money over their three unique stocks in the same way: 25% of their money in their Stock A, 55% in their Stock B, and 20% in their Stock C. The following table gives the combined increases on each total investment derived from the respective increases in specific stocks:

Name	x (Stock A)	y (Stock B)	z (Stock C)	Effective % Return = $0.25x + 0.55y + 0.20z$
Hypatia	9.6%	9.5%	9.9%	9.605%
Al-Samawal	9.0%	9.2%	9.4%	9.19%
Dardi	8.8%	9.0%	9.3%	9.01%
Ramus	9.6%	8.8%	5.4%	8.32%
Franklin	3.0%	5.0%	9.5%	5.4%
Darwin	7.8%	6.8%	7.2%	7.13%

For example, if Ramus' stocks increased in value by 9.6%, 8.8%, and 5.4%, respectively, then the money that he invested in all three has a combined return of 8.32%. This means that we could think of the three individual investments working together as one big single investment that yielded an 8.32% return. As such, if Ramus invested a total of $50,000 in his three-stock portfolio, then his money would have increased by 50000(0.0832) = $4160—more than twice as much as the increase in the safe 4% investment. Note that our initial investor's circumstances with a 5.4% overall increase match Franklin's in the table.

It is also possible for a stock's value to decrease. When this happens, we can use the same formula and simply introduce negative numbers to represent the losses. For example, if Dardi kept his investment percentages the same but his portfolio performance changed such that Stock A increased in value by 6.7%, Stock B decreased by 6.2%, and Stock C decreased by 3.5%, his return could be represented as follows:

x (Stock A)	y (Stock B)	z (Stock C)	Effective % Return = $0.25x + 0.55y + 0.20z$
6.7%	−6.2%	−3.5%	$0.25(6.7\%) + 0.55(-6.2\%) + 0.20(-3.5\%) = -2.435\%$

This means that if he invested $50,000, then his money would have changed in value by 50000(−0.02435) = −$1217.50, which represents a loss of $1217.50. This, of course, would be a worse situation than placing his money in the safe investment with a guaranteed 4% increase on his money.

Dardi is free to forecast other values of x, y, and z while keeping his investment percentages the same, thereby leaving the parameters alone. Alternatively, he could change the scenario entirely by distributing his money differently among the three stocks—changing investment percentages a, b, and c—or by diversifying even further by distributing his money over five stocks instead of three, which would mean the addition of two more parameters and two more variables, as happened with the GPA formulas.

In the next section, we briefly discuss how mathematicians handle situations that require them to deal with larger numbers of variables.

SUBSCRIPTED VARIABLES

When we calculated GPAs, we were confronted with a situation that had more variables than our traditional variables x, y, and z could cover. This meant that we had to introduce more letters to handle them—u and v. However, there were only two additional variables to contend with in that case. What should we do if we have 10, 15, or even 30 variables? Clearly, assigning variables according to Descartes' method of using later letters in the alphabet for variables and those early in the alphabet for parameters will founder under such an explosion.[4]

These situations can and do happen, and mathematicians throughout history have come up with creative protocols to deal with them. The medieval Indians on the subcontinent came up with a color scheme to handle situations involving multiple unknowns, although parameters as we know them today appear not to have existed for them. Their primary unknown was *yāvattāvat*, or "as much as so much," which approximates the x that we use today. When they encountered situations involving several unknowns, they innovated by using colors. Describing their scheme in the 1100s, Indian mathematician Bhaskara II wrote:

> [One] unknown (yāvattāvat) is the color black, another blue, yellow, and red. [Colors] beginning with these have been imagined by the best of teachers to be the designations of the measures of the unknowns in order to accomplish their calculation.[5]

In order to further simplify matters, they frequently abbreviated to only the first syllable. In the case of *yāvattāvat*, they would use *yā*.[6]

In modern times, mathematicians have developed other resourceful ways to handle multiple variables and parameters, usually injecting subscripted numerals into the notational scheme. We will discuss one such scheme that solves the problem at hand while maintaining consistency with the Descartes protocol.

Let's consider the case where we have five variables. We decided earlier to choose the letters x, y, z, u, and v to represent them, but using this new notation, we can instead use a subscripted format of $x_1, x_2, x_3,$

x_4, and x_5. Although we repeat the use of x for each new variable, this does not imply a derivative relationship between these variables. We can think of the subscripts as telling us that x_1 represents the first variable instead of x, x_2 represents the second variable instead of y, and so on up to x_5 representing the fifth variable instead of v. We can replace parameters a, b, c, d, and e in a similar fashion with the subscripted format a_1, a_2, a_3, a_4, and a_5.

This means that the expression for GPA translates in the following way:

$$ax + by + cz + du + ev \rightarrow a_1x_1 + a_2x_2 + a_3x_3 + a_4x_4 + a_5x_5.$$

Note that this subscripted formula possesses a convenience that the un-subscripted one does not. For instance, if you were asked with which variable the parameter d was associated, you might have to go back and do a bit of comparing to see that the answer is u. However, when utilizing subscripts, if you were asked with which variable the parameter a_4 was associated, you could immediately say x_4. If we simplify the GPA expression to $4x_1 + 3x_2 + 2x_3 + 1x_4 + 0x_5$—where $a_1 = 4$, $a_2 = 3$, $a_3 = 2$, $a_4 = 1$, and $a_5 = 0$—the fall semester GPA table would become the following:

Variables	Total Credits	Percentage (Out of 20 Credits)
x_1 = total A grade credits	7	35%
x_2 = total B grade credits	5	25%
x_3 = total C grade credits	4	20%
x_4 = total D grade credits	0	0%
x_5 = total F grade credits	4	20%

Fall semester credit totals

Performing the calculations using the relabeled formula will still yield the same GPA of 2.55.

We can see how this new notational system can scale to situations involving even more variables—for instance, in a case involving 10 variables, we would simply use the 10 subscripted variables $x_1, x_2, x_3, \ldots,$ x_8, x_9, x_{10}.

A FEW MORE EXAMPLES

There are many more examples of quantitative cocktails beyond those we have discussed here. Like our GPA calculation, some of them can be simplified, doing one grand division at the end of the calculation, to avoid rounding decimals multiple times in the process. However, here we will discuss them from the point of view of the subscripted expression: $a_1x_1 + a_2x_2 + a_3x_3 + a_4x_4 + a_5x_5$.

SURVEY SCALES

If you've taken a survey recently, you may immediately recognize the potential for variation within them. Respondents are given a series of questions and are expected to answer them according to a rating scale. Many surveys use psychometric rating scales, often called Likert scales after their inventor, Wyoming-born Rensis Likert. Some scales use the following ratings and number assignments: 1 = very unsatisfied, 2 = unsatisfied, 3 = neutral, 4 = satisfied, 5 = very satisfied, or 1 = strongly disagree, 2 = disagree, 3 = neutral, 4 = agree, 5 = strongly agree. Then, for any given question, the individual answers from all respondents are blended into a collective result.

	Survey Question Score Cocktail 1	Survey Question Score Cocktail 2	Variables
Ingredient 1	% of answers receiving very unsatisfied	% of answers receiving strongly disagree	x_1
Ingredient 2	% of answers receiving unsatisfied	% of answers receiving disagree	x_2
Ingredient 3	% of answers receiving neutral	% of answers receiving neutral	x_3
Ingredient 4	% of answers receiving satisfied	% of answers receiving agree	x_4
Ingredient 5	% of answers receiving very satisfied	% of answers receiving strongly agree	x_5

Using the parameter assignments $a_1 = 1$, $a_2 = 2$, $a_3 = 3$, $a_4 = 4$, and $a_5 = 5$, the formula for calculating the blended scores for each question $a_1x_1 + a_2x_2 + a_3x_3 + a_4x_4 + a_5x_5$ becomes

$$1x_1 + 2x_2 + 3x_3 + 4x_4 + 5x_5.$$

For example, if 5 respondents answer Question 7 on the survey with very unsatisfied, 6 answer unsatisfied, 8 answer neutral, 21 answer satisfied, and 10 answer very satisfied, we can calculate the following percentages out of 50 responses: $x_1 = 10\%$, $x_2 = 12\%$, $x_3 = 16\%$, $x_4 = 42\%$, and $x_5 = 20\%$. This will yield an overall score for Question 7 of $1(0.10) + 2(0.12) + 3(0.16) + 4(0.42) + 5(0.20) = 3.5$, meaning that the collective feeling on this question is halfway between neutral and satisfied.

Recalling course averages, we can think of all the individual survey responses mixing together to yield a result that would be equivalent to a circumstance where each respondent selected 3.5. Of course, in the context of the problem, possible responses can only be whole numbers, so we can also think about this as an even split of respondents scoring 3 and 4. The other questions would be scored in the same fashion.

This procedure can be applied to many rating systems, including those that use stars to evaluate various products and services such as restaurants, hotels, books, or films. Some such rating systems, however, may employ additional methodology besides what has been discussed here to generate their final ratings.

SURVEY CATEGORIES

Consider the results of a survey of 1000 households to determine the average number of televisions they have in the home:

Number of TVs per Household	Number of Households
0	40
1	90
2	460
3	310
4	100
Total Households:	1000

We could summarize the overall result of all of these responses using a similar formula $a_0x_0 + a_1x_1 + a_2x_2 + a_3x_3 + a_4x_4$ with a slight adjustment, this time with parameters $a_0 = 0$, $a_1 = 1$, $a_2 = 2$, $a_3 = 3$, and $a_4 = 4$ and variables x_0 = percentage of households with zero TVs, x_1 = percentage of households with one TV, and so on up to x_4 = percentage of households with four TVs. You'll notice that in this case we've chosen to begin with a_0 and x_0 rather than a_1 and x_1 so that the parameter and variable subscripts match the number of TVs.

Placing the values in for the parameters yields the expression $0x_0 + 1x_1 + 2x_2 + 3x_3 + 4x_4$. The percentage of households in each category of TV ownership are $x_0 = 4\%$, $x_1 = 9\%$, $x_2 = 46\%$, $x_3 = 31\%$, and $x_4 = 10\%$. Substituting these values into our general formula yields $0(0.04) + 1(0.09) + 2(0.46) + 3(0.31) + 4(0.10) = 2.34$. This means that, for the 1000 households surveyed, the average one has about 2.3 TVs in the home, or every three households have about seven TVs between them.

BASEBALL SLUGGING PERCENTAGE

The official website of Major League Baseball defines slugging percentage as follows:

> Slugging percentage represents the total number of bases a player records per at-bat.…Slugging percentage differs from batting average in that all hits are not valued equally. While batting average is calculated by dividing the total number of hits by the total number of at-bats [note that a walk does not count as an at-bat], the formula for slugging percentage is: $(1B + 2Bx2 + 3Bx3 + HRx4)/AB$.[7]

A player's slugging percentage could just as easily be called their slugging average. The idea behind a slugging percentage is that it differentiates between the type of hits a batter gets. Thus, a player who gets more extra base hits will generally fare better with slugging percentage than a player who hits only singles. The traditional batting average does not distinguish between the different types of hits, meaning that a single and a home run have the same value in the calculation.

The calculation for slugging percentage is operationally similar to a GPA; however, there are some key differences which include the following:

- The values for slugging percentage are given in thousandths, so where a GPA would be given as 0.52 the equivalent slugging percentage would be given as 0.520 (pronounced 520).
- Slugging percentages for very good baseball players are far lower than GPAs for very good students. That is, though it is not unusual for very good students to get straight A's (or a 4.0 GPA), good players don't get "straight home runs." The highest slugging percentage in major league baseball history for a single season is 0.863 by Barry Bonds in 2001.[8] A 0.863 GPA is below a straight D average.

If we were to convert slugging percentage into a quantitative cocktail, we would have the following:

	Slugging Percentage Cocktail	**Variables**
Ingredient 1	% of at-bats that are outs	x_1
Ingredient 2	% of at-bats that are singles	x_2
Ingredient 3	% of at-bats that are doubles	x_3
Ingredient 4	% of at-bats that are triples	x_4
Ingredient 5	% of at-bats that are home runs	x_5

The parameter values would therefore be $a_1 = 0$, $a_2 = 1$, $a_3 = 2$, $a_4 = 3$, and $a_5 = 4$, and applying them to our general formula will yield

$$\text{slugging percentage} = 0x_1 + 1x_2 + 2x_3 + 3x_4 + 4x_5.$$

Note that the formula on the MLB website leaves out the 0 value term in the numerator for when an out occurs. This is not a problem because multiplying by 0 would make this term disappear anyway. The outs made still get recorded in the total number of at-bats in the denominator.

The possibility of relabeling this formula starting at x_0 and ending at x_4, as we did in "Survey Categories," remains an option here as well.

Let's see how the slugging percentage plays out in the following 25 at-bat scenario where a player recorded 16 outs, 5 singles, 2 doubles, 1 triple, and 1 home run, adding up to 16 total bases. Out of 25 at-bats, this would give our variables the following values: $x_1 = 64\%$, $x_2 = 20\%$, $x_3 = 8\%$, $x_4 = 4\%$, and $x_5 = 4\%$. The slugging percentage formula therefore becomes

$$0(0.64) + 1(0.20) + 2(0.08) + 3(0.04) + 4(0.04) = 0 + 0.2$$
$$+ 0.16 + 0.12 + 0.16 = 0.64.$$

We would interpret this as a slugging percentage of 0.640, which would be considered an extremely good slugging percentage. To put it in perspective, this means that this batter on average gets 0.64 bases for every at-bat. We can check this by multiplying the 25 at-bats by 0.64, which yields 16 total bases out of 25 at-bats, which matches the scenario. Note that the straight batting average $\dfrac{\text{number of hits}}{\text{number of at-bats}}$ would be given by $\dfrac{9}{25}$ or 0.360.

Another way to view what the slugging percentage is telling us is to divide the slugging percentage by the batting average. This will give us the number of bases per hit that the player gets on average. Here that would be $\dfrac{0.640}{0.360}$, or approximately 1.78 bases per hit. You can also check this directly by dividing the 16 bases by the 9 hits. This means that we can think of this batter as getting close to a double for every hit.

The ratio of number of bases per hit by itself is not as useful as the slugging percentage because it doesn't take outs into account. For instance, a player who gets 1 hit out of every 50 at-bats, but that hit is a home run, will have a ratio of 4 bases per hit, and yet they'll get out 98% of the time. The batter's slugging percentage, which does account for outs, would be a very low 0.080, or 0.08 bases per at-bat.

ENHANCED REASONING

Back in Chapter 1, we discussed the hypothetical situation of a child discovering new words through rhymes rather than synonyms. The former ties words together based on having similar sounds, whereas the latter method ties together words based on common meanings. Both turn out to be useful and systematic ways to learn and remember new words.

Similar to that child, in this chapter we have tied together diverse numerical phenomena that share common algebraic and conceptual features in order to learn more about them. We have done this before in the classroom word problems of Chapter 5, where phenomena were connected by similar types of equations, and in Chapter 7, where five classroom and real-world problems were linked together around an identical simplified equation. Linking together conceptually similar entities is something that nearly all of us do, with the process forming an important foundation to analogical and metaphorical thinking more generally. What distinguishes and intensifies what we have been doing here is the way that we have incorporated algebra into our thinking.

The Spanish philosopher José Ortega y Gasset described it thus:

> The metaphor is perhaps one of man's most fruitful potentialities. Its efficacy verges on magic, and it seems a tool for creation which God forgot inside one of His creatures when He made him.... The metaphor alone furnishes an escape; between the real things, it lets emerge imaginary reefs, a crop of floating islands.[9]

What algebra does par excellence is sharpen these tools to a much finer point. Put another way, algebra singularly weaponizes metaphorical and analogical reasoning, rendering it more precise and operational. This is true of mathematics in general, but algebra forms one of the strongest alloys used to forge this mighty mathematical sword.

We have employed this enhanced reasoning twice in this chapter, first by tying together some types of table and spreadsheet analysis with algebraic reasoning and second by examining different variations on the theme of quantitative cocktails. What useful knowledge has been gained by doing this?

In Chapter 6 we discussed, as part of the algebraic experience, how algebraic thinking can help us gain more insight into numerically varying phenomena while also giving us greater technical confidence and comfort in dealing with such situations. The goal of this chapter has been to see this premise in action, as well as highlight the emerging possibilities for deeper understanding and new connections that become possible through the strategic deployment of algebraic technique.

Let's revisit some of the major conceptual takeaways from this chapter that emphasize the strengths of algebraic thinking.

EXPRESSING VARIATION WITH TABLES AND SPREADSHEETS

By connecting certain aspects of tables and spreadsheets to algebraic expressions, we identified a conduit between these two important ways of expressing numerical information. Moreover, the use of specific examples allowed this connection to be exploited in a bit more detail and depth. The examples we analyzed are only a few of the hundreds and thousands possible—some relevant to everyday life and some, like the number of days and age problem, more gimmicky than practical.

Although the number of days and age problem may have no direct bearing on everyday life, the relationships, properties, and strategies essential to understanding the problem—like symbolically and efficiently organizing ensembles of numbers—can and do have direct applicability to a wide range of circumstances. The analysis of data in tables and spreadsheets will inevitably come up time and again for most of us, and so the hope is that by highlighting the relationship between algebra, tables, and spreadsheets, we can strengthen our understanding of all three tools and their interaction.

QUANTITATIVE COCKTAILS

By exploring problems that blend distinct, unequal portions together into a whole, we have uncovered additional conceptual treasures.

PRECISION/INTUITIVE INTERPLAY: Algebra helps us operationalize and control our mixing of diverse quantities with mathematical precision. Although our general ability to recognize patterns makes it possible for us to intuit connections between affected phenomena, it is algebra—with its vocabulary of variables and parameters—that gives us a language in which to explicitly identify the substance of that connection. Qualitative metaphors like cocktail mixing also strengthen conceptual comprehension. The precise and qualitative features of this enhanced reasoning can work together in powerful, almost magical conjunction.

ALGEBRAIC SUPERHIGHWAY: When we use algebraic language to express connections and establish relationships, we are able to link seemingly isolated phenomena into a powerful symbolic superhighway. Much like an interstate connecting isolated towns can enhance commerce and exchange within that network, so too can our understanding of diverse phenomena be greatly strengthened by algebra.

We tried to do this explicitly in the case of calculating GPAs by illustrating its similarities to calculating course averages, in hopes that we could leverage our existing understanding to better conceptualize the new circumstances. Mathematicians and scientists do this on a regular basis, using words such as *realizations*, *representations*, *isomorphisms*, *homeomorphisms*, and *canonical forms* to describe the sophisticated linking together of diverse ideas, objects, and behaviors for conceptual insight and computational advantage.

THE TRADITIONAL ARITHMETIC AVERAGE HAS SIBLINGS: Calculating more sophisticated averages has helped establish a more contextual view of how they relate to the traditional average and what that comparison tells us.

For instance, to find the average of the three numbers 30, 50, and 95, we would calculate $\dfrac{30+50+95}{3}$ to arrive at $58\dfrac{1}{3}$, rounding to 58. However, by performing a symbolic maneuver, we could rethink this division as

$$\frac{1}{3}(30)+\frac{1}{3}(50)+\frac{1}{3}(95)=58\frac{1}{3}.$$

These three numbers correspond to Franklin's scores in the three assessment categories of homework, tests, and the final from the section on computing grades.

This maneuver now shows us that in the formula $ax + by + cz$, we have the parameters a, b, and c each equal to $\dfrac{1}{3}$ or $33\dfrac{1}{3}\%$, which yields

$\frac{1}{3}x + \frac{1}{3}y + \frac{1}{3}z$. The variables x, y, and z retain their meaning, which here makes them equal to 30, 50, and 95, respectively.

Thus, the traditional average itself is a quantitative cocktail corresponding to the situation where each category contributes the same amount to the total. In the case of three numbers, each contributes $33\frac{1}{3}\%$ to the total. If there were five numbers—or categories—each number would contribute $\frac{1}{5}$ or 20% to the total, and so on.[10]

The traditional average is generally determined from a single division for the same reasons that one grand, final division is preferable when calculating GPAs—we obtain a more accurate result. Thus, rather than standing alone as an isolated entity, the traditional average turns out to be one of an entire family of different types of averages. This family of averages is more commonly called *weighted averages*, a conceptually important mathematical category in its own right.

GATEWAY TO NEW IDEAS: All versions of an idea are not created equal. Identifying the effective percentage rate of returns on stock investments as a quantitative cocktail opened up a new situation—the possibility that the impact of one category can take away from the impact of another category and that effects can be subtractive. In the first three examples, the impact from the various categories all reinforced one another.

A CHILD SEARCHING FOR NEW WORDS: Though we discovered that effects from a given category can be subtractive as well as additive through a specific example, we could have inferred as much directly from the formula $ax + by + cz$ itself. That is, nothing prevented us from asking the question: Can we use this formula if one or more of the products—ax, by, or cz—are negative?

This question may have led to examples that fit the scenario in question, which may in turn have pointed us to "effective percentage rates of returns." This type of critical inquiry and mathematical probing can and does occur in the higher levels of the subject and is a useful catalyst for generating new ideas and discoveries, just as an exploratory approach to language produces new words and opens new possibilities for a child.

COMPUTATIONAL GAIN: When we set out to purchase a new piece of technology such as a digital camera or a computer, we have predetermined criteria in mind that form the basis of our decision-making, some more important to us than others. For a computer, these various properties may be the type of microprocessor, the speed of the microprocessor, RAM, hard-drive capacity, or price. For a digital camera, they could be aspects like image resolution, battery life, ISO range, cost, or a built-in flash.

For the consumer, these categories can help them assess many individual devices, just as homework, tests, and the final exam enable teachers to assess their students. Like a teacher, the consumer can assign contribution percentage values—or weights—to each of these categories or properties based on their relative importance to them, giving each device their own personal "grades." In the case of a digital camera, some consumers will place a higher premium on image resolution over ISO range, and so on. A built-in flash may not matter at all to certain consumers, leading them to dismiss that property from consideration just like the instructor who doesn't factor attendance into a student's final grade. In some cases, the scores for the camera categories will be on a scale from 0 to 5 or 0 to 10 or 0 to 100, and will be rated by an expert in a magazine or online.

Let's say, for example, that a given consumer ranks categories of importance for their digital camera purchase as follows: 50% for image resolution, 20% for battery life, and 30% for ISO range. The formula to rate prospective cameras will then be 0.50(image resolution score) + 0.20(battery life score) + 0.30(ISO range score). In our standardized language, this would establish parameters $a = 0.50$, $b = 0.20$, and $c = 0.30$ with variables x = image resolution score, y = battery life score, and z = ISO range score—resulting in the expression $0.50x + 0.20y + 0.30z$.

Now, let's assume that a magazine rates a particular model of camera at a 6 for image resolution, an 8 for battery life, and a 7 for ISO range on a 10-point scale. The overall score for this camera on the consumer's grading scale would therefore be 0.50(6) + 0.20(8) + 0.30(7) = 3 + 1.6 + 2.1 = 6.7. In a similar fashion, the customer could give personal grades to other cameras and then make their determination by comparing those values. Of course, just like instructors grading students, other customers will weigh the importance of each category differently, resulting both in different parameter values and in different scores or grades for the cameras.

Computers, cars, houses, and even colleges can be scored and compared this way. It's worth remembering that algebra doesn't remove subjectivity and shouldn't remove common sense from the decision-making process—emotions, gut instincts, and other intangibles still matter—but it may help to impose a bit of consistency, organization, and clarity on what may otherwise seem like a tidal wave of options.

CONCEPTUAL GAIN: A quantitative cocktail that affects multiple tens of thousands of students every year is the well-known *U.S. News & World Report* College Rankings. The following table lists the 2021 categories and their weights for universities and liberal arts colleges[11]:

	U.S. News & World Report National Universities and Liberal Arts Colleges Cocktail	Variables
Ingredient 1	Graduation and retention rates (**22%**)	x_1
Ingredient 2	Social mobility (**5%**)	x_2
Ingredient 3	Graduation rate performance (**8%**)	x_3
Ingredient 4	Undergraduate academic reputation (**20%**)	x_4
Ingredient 5	Faculty resources (**20%**)	x_5
Ingredient 6	Student selectivity (**7%**)	x_6
Ingredient 7	Financial resources (**10%**)	x_7
Ingredient 8	Average alumni giving rate (**3%**)	x_8
Ingredient 9	Graduate indebtedness (**5%**)	x_9

The parameter assignments are respectively

$$a_1 = 0.22, a_2 = 0.05, a_3 = 0.08, a_4 = 0.20, a_5 = 0.20, a_6 = 0.07,$$
$$a_7 = 0.10, a_8 = 0.03, a_9 = 0.05.$$

Plugging these values in yields a University and College Ranking Score:

$$0.22x_1 + 0.05x_2 + 0.08x_3 + 0.20x_4 + 0.20x_5 + 0.07x_6 + 0.10x_7 + 0.03x_8 + 0.05x_9.$$

Once scores for each of the variables are known for a given school, we can plug them into this formula and complete the calculation. However, the *U.S. News & World Report* applies a bit of additional methodology to obtain the final scores that they publish.

Now if you are so inclined, you have the tools to change the parameter assignments for each of the various categories and come up with your own personal ranking system for colleges and universities based on the relative importance of each category to you—thus no longer being a spectator in this particular area, but a player with the agency to interact with and manipulate algebra for your own needs.[12]

CONCLUSION

The narratives as discussed in this chapter would be hard pressed to enter our consciousness without some awareness of algebra. Crucial to it all is the recognition that when we encounter algebraic ideas outside of the classroom, they are usually cloaked in other garb. They rarely come at us in the x's, y's, and z's of school algebra.

The critical "tell" of algebraic possibilities in the wild is the presence of numerical variation of some sort. Sometimes what we are interested in may only be a single instance of a variable situation. The objects that generate those variations can be individual people distinguished by their performances in a class, grades in several classes, income, survey responses, or batting results; or they could be variations in company stock prices, ratings of different electronic objects, and ratings of different colleges and universities, as well as a host of other possibilities.

Once it has been established that variation is present in a situation, either by personal investigation or more likely through formulas obtained elsewhere—the internet, by word of mouth, or a book—it sometimes becomes possible to replace explanations—often given in words, via strange symbols, via visual diagrams, or as tables of values—by the familiar x's, y's, and z's, or subscripted symbols and parameters, too, if needed.

This standard notation offers a valuable orienting principle, as a landmark does, to understand and interact with numerical phenomena. Sometimes the abbreviations in their original form are powerful enough symbols in their own right, or are so common that we tend not to retag the variable, as with $E = mc^2$. Either way, algebra is there to play some role in helping us to better understand and interpret relevant quantitative problems—but only if we want it to. It is reminiscent of the possibilities afforded us by the panorama of history to better understand and interpret our own historical times, where we have powerful tools at our disposal that offer us insight to make better, more informed decisions for the future.

But for individuals to take advantage of the warmth and brightness from either of these conceptual lights, we must first know where to look, appreciate their power to enlighten, learn how they work, and, ultimately, have the courage to flip the switch and turn them on.

9

Algebraic Flights: Mechanism and Classification

> The true method of discovery is like the flight of an aeroplane. It starts from the ground of particular observation; it makes a flight in the thin air of imaginative generalization; and it again lands for renewed observation rendered acute by rational interpretation.
>
> —Alfred North Whitehead (1861–1947), *Process and Reality: An Essay in Cosmology*

We continue on in the spirit of the last chapter, briefly turning our attention now to what we have termed the second of the great algebraic dramas—the art and science of describing situations by equations and then finding their solutions for insight.

An instructor in a class of 24 asks for a volunteer to pick a whole number between 1 and 24 and to tell it to the rest of the class after the instructor is out of earshot. They exit the classroom, close the door, and take a stroll down the hall before returning. Once back, the instructor directs everyone to take the chosen number (whatever it is) and multiply it by 15, then to add 20. Someone says they got 125, which the rest of the class immediately confirms.

The instructor then goes behind the projection screen making sounds as if they are frantically looking for something and then comes back out stating that they have found the original number chosen by the volunteer, and that it is 7. Some students chuckle, a few seem genuinely amazed, but most smile knowingly.

One of them blurts out, "that's easy, you just reversed the steps." The teacher nods in agreement and shows what happened in algebraic language:

	Steps of the Procedure	Algebraic Interpretation at Each Step
Step 1	Class picks a number whose value is unknown to teacher	x
Step 2	Class multiplies this number by 15	$15x$
Step 3	Class then adds 20 to the result	$15x + 20$
Step 4	Class obtains 125	$15x + 20 = 125$
Step 5	Teacher solves the equation in their head	$x = 7$
Step 6	Teacher reveals that the chosen number is 7	

The reduction diagram shows the systematic solution process:

This is a common type of exercise in an elementary algebra class, and though simple once understood, it still contains powerful conceptual fuel.

What understanding algebraic mechanisms has done here is allow the instructor to find out a piece of information that they didn't directly know. At no point in the process of discovery did they ever hear any of the students say the number 7. Yet, after being given the value 125 and based on the way certain algebraic relationships work, they were able to symbolically link this number to the value that they didn't know, and then use canonical maneuvers to uncover what they wanted; namely, that the student had chosen the number 7.

MECHANISMS

It turns out that familiarity with the mechanisms of phenomena—or at least of their effects—can aid us in finding out unknown information in a wide variety of areas.

Consider lake levels. A lake is an interconnected entity where if the water level in one place rises, there will be corresponding effects elsewhere. This will allow us to deduce, by proxy, information about the lake and its surroundings without necessarily seeing those locations. For instance, if the lake level in our location rose to 6 feet above normal, then we will know that a spot on the shore, thirty miles away and only half a foot above the lake during normal levels, will now be underwater.

Here, we used an understanding of the way water levels work on Earth to find out an unknown piece of information. In this case only information from a local measurement was used; no calculations were needed.

Consider river currents. A boy accidentally drops a baseball bat off of a bridge into the center of a rapid-less river whose current is 3 miles per hour along its entire length. If a rafter 15 miles downstream pulls the bat out of the center of the river at 1:00 PM, when did the boy drop the bat into the river?

To solve this problem, we could use the formula distance = (speed) times (the time of travel), or $d = st$. Here, for convenience and since the terms are simple enough to remember, the variables have been abbreviated to their first letter as opposed to using generic x, y, and z. This follows common usage for the distance formula, although "rate"— abbreviated to r—is often used in the place of speed.

Given that $s = 3$ and $d = 15$, the formula simplifies to $15 = 3t$, which immediately gives $t = 5$. Thus, the approximate time it took for the bat to travel to where the rafter picked it up was 5 hours. This means that it was dropped into the water around 8:00 AM. In this case, two calculations were used to obtain the desired information.

The values here are basic enough that many could quickly compute the answer without resorting to formal algebra at all, but we are after far bigger conceptual game than simply finding a quick solution to this basic problem, and using algebra gives us the tools to set the stage.

LITERAL EXPRESSIONS

Though the previous presentation of finding the time that the bat was dropped into the river certainly involves familiar techniques, the perspective is subtly different from the viewpoint in most of the previous chapters. Thus, before we take our first flight, we will take a brief detour to see how the perspectives compare.

NAMING ALGEBRAIC EXPRESSIONS

Throughout most of the book, we have not used proper names to describe algebraic expressions at all. For instance, the variable expression that encoded the procedure for the number of days and age problem $(100x + 2013 + z - y)$ was never given a formal name.

It is only in the previous chapter that we have consistently given proper names to particular recurring notions: course average, final exam average, GPA formula, effective percentage rate of return, slugging percentage, and University and College Ranking Score.

When proper names are given in this way, the expressions are sometimes called *formulas*.

BOTH SIDES OF FORMULA ARE NOW IN PLAY

Though proper names were given to the formulas in the previous chapter, we still didn't abbreviate them to a single letter. For instance, we had the following:

- Course average = $0.25x + 0.55y + 0.20z$.
- GPA = $4x + 3y + 2z + 1u + 0v$.
- University and College Ranking Score = $0.22x_1 + 0.05x_2 + 0.08x_3 + 0.20x_4 + 0.20x_5 + 0.07x_6 + 0.10x_7 + 0.03x_8 + 0.05x_9$.

Notice that the names of the formulas on the left-hand side are not abbreviated to a single letter. If we wanted to be consistent in this chapter, we should have written the distance formula as distance = st. The fact that it is often the practice to also abbreviate "distance" to the single letter d in this formula—though not necessarily standard practice before in situations like those in Chapter 8—is a nuance worth noting.

The nuance is that in the previous chapter we didn't view the proper-named term on the left-hand side of the equals sign in the same way that we viewed the variable terms on the right. The variables and parameters on the right-hand side of the formula were viewed in a dynamic way, meaning we could substitute in different numbers for them and then calculate. Once done, this would give us the value of the proper-named term on the left-hand side (GPA, course average, etc.).

The proper-named terms were treated as non-interacting objects whose values were totally dependent upon the values on the right-hand side. We viewed them as if they inherited their variability—their very existence even—completely from the expressed combination of variables and parameters on the right. For example, GPA doesn't exist without the combination of grade point values and credit percentages that make it up.

In this chapter, we are now viewing the distance in the same dynamic and measurable way as the entities on the right-hand side—speed and time. We see the distance as something that independently exists without being subordinate to the other entities. Consequently, the distance is viewed more as a quantity that is definitely related to the variables—s and t—on the right-hand side in the case of motion, but not as something whose very existence depends on them.

The conversion formula, which connects the independently designed yet related Fahrenheit and Celsius temperature scales, provides an example of this point. Fahrenheit and Celsius are measures of temperature that can be related to each other through the formula $F = \dfrac{9}{5}C + 32$. Yet each of these scales stands on their own two feet, independent of the other. Thus, the formula highlights a relationship between two independent entities rather than demonstrating a subordinate dependence of one quantity upon the other.

In fact, in the example we gave, the distance and speed were assigned numbers while we treated the time as an unknown that was related to them. In this viewpoint, terms on both sides of the equation are in play. And now because the distance, too, is looked at as being interactive and up for maneuver, it is not uncommon to abbreviate it to a single letter.

This is more of a subtle practice in some educational contexts, however, and not a necessary or absolute requirement by any means. That is, in any of the three formulas in the previous list, it is certainly possible to put the terms on the left-hand side of the equation in play, but for many uses the standard practice is not to do so.

The interplay between using the functional notation $f(x)$ versus the letter notation y also touches upon these subtleties. See Appendix 3 for a brief discussion.

THE SYMBOLS ARE DESCRIPTIVE ABBREVIATIONS

In previous chapters, the x's and y's (a's and b's) we predominately used to represent variables (parameters) were standard characters that gave no clue on their own as to the quantities they were representing.

As mentioned in Chapter 1, there are advantages to using these generic representations, in that they give us a standard font in which to represent and compare unfamiliar phenomena. This familiarity of representation serves a valuable role both in educating the novice and in more easily assisting the uncloaking of unifying connections between diverse behaviors. The latter was copiously on display in the previous chapter.

However, in the distance formula, the symbols that are used do give clues as to the quantities that they represent—they are used in an interpretive sense. This type of representation becomes the predominant way of doing business in science, finance, and other real-world applications. Equations of this type are often called *literal equations* in algebra textbooks.

More generally, the terms *literal equation* and *literal expression* are used to describe equations or expressions that have parameters mixed in with variables or when multiple variables alone are involved, regardless of whether or not they involve descriptive abbreviations, such as $E = mc^2$, or generic ones, such as $ax + by + cz$. They roughly approximate the type of equations and expressions that we have termed "big algebra" in this book. And though we have so far only hinted at it in a few places, these literal expressions and equations can also be maneuvered for major conceptual and computational gain wherever they appear. Many people find great difficulty, unfortunately, with such expressions in elementary algebra, often viewing them as a confusing maze of letters.

PARAMETERS DISGUISED?

When we decide to employ descriptive symbols into expressions and equations, this can start to clash with some of the protocols that we have previously used for alphabetic symbols: the Descartes protocol, for representing variables and parameters, in particular. For example, the right-hand side of the expression $d = st$ has the same form as the multiplied terms on the right-hand side in the previous chapter (such as $ax + by + cz$), but in each of those three terms (ax, by, and cz), we always identified one letter as the variable and one as the parameter by assigning them a letter from a specific location in the alphabet.

For the case here (st), the letters are descriptive and were obtained by simply abbreviating the names of the phenomena they represent, meaning that their location in the alphabet is the result of happenstance. Consequently, the location of a letter can't be counted on to tell us anything of algebraic significance here, thus the Descartes protocol may get completely blown out of the water for this type of situation.

In spite of all of this, it is still the speed, s, here that more naturally segments the phenomena into individual scenarios like a parameter does. That is, an individual river defines a given scenario, and for a particular location and date on the river, it is the speed that acts more like the constant, whereas the time spent on the river can vary. This allows us to think of the hundreds of rivers as each having a certain respective speed, at least in the locations of interest.

The s can then be used to algebraically tune the formula $d = st$ to describe the scenario of floating down locations on a particular river. For example, given three rivers with current speeds of 3 mph, 2.8 mph, and 6.5 mph, we would tune s to 3, 2.8, or 6.5—yielding the three expressions $d = 3t$, $d = 2.8t$, and $d = 6.5t$, respectively.

This becomes even more apparent when we consider other phenomena that have more well-defined speeds throughout the time of travel than lengthy rivers, where more often than not the volume of flow per unit time is a more useful measure than the speed of the current. For instance, a person traveling in a car at the constant speed of 50 mph on an open stretch of freeway would define a scenario given by $d =$

50t, while passengers in an airplane cruising at the constant speed of 450 mph would define yet another one described by $d = 450t$, and so on.

It is the distance and time of travel that more naturally can change within each of these scenarios, at least, as we have described them. So, though it wasn't necessary for solving the equation involving the bat, we can still think of speed as the parameter, and time and distance as the regular variables. We can then classify various phenomena with this organization in mind if we'd like.[1]

If desired, we could indicate the parameter nature of the speed by partially returning to the Descartes protocol and rewriting the expression as $d = ax$, where d = distance, a = speed of the river, and x = time on the river.

In courses today, you may sometimes see these situations represented by the formula $y = mx$, where y = distance, m = speed, and x = time. Though not early in the alphabet, the m is still viewed as a parameter corresponding geometrically to what is called the slope of a line, whereas the x and y are both looked upon as being changeable within the scenario—often a graphical one—and thus as the regular variables.[2]

CONCEPTUAL FLIGHT

We have seen that motion at a constant speed in water, on land, and in the air can be described by the same types of expressions. The mechanisms that propel the objects in each of these three media are different: the river for the bat, internal combustion engine or electric motor for the car, and jet engine for the plane. But the differences in mechanism—and whether or not we even deeply understand them—is irrelevant to the algebra that describes them.

As long as we have the knowledge to symbolically describe how the location of the object changes with time and a few other key facts, such as the distance between the starting and ending positions in these cases, then we can use the formula $d = st$ to solve for t and hence find the time of travel.

Curiously enough, on a more general level, this parallels in a way the situation in which the teacher is engaged when finding the number chosen by the student. For the classroom problem, there was no physical motion at all, only two instructions, yet those instructions did something mathematically similar to what the river and the car did. They connected two numbers via a mechanism, the number (x) chosen by the student—but unknown to the teacher—to the calculated number (125) shared with the teacher. If you change one, then the other will change to keep the relationship describing the mechanism valid and vice versa.

In a similar fashion, if the speed is known, then the motions in the river, the car, and the plane can be seen as connecting two numbers— the distance traveled and the time of travel. If one quantity changes, then so must the other to keep the relationship valid. This means that on a certain level these motions can be thought of in the same way as the teacher's set of two instructions.

If we want to represent these connections in a way that lends itself to clarity and maneuver, then employing algebraic equations is a useful way to do it. The following table illustrates this:

Object	Unknown Number	Mechanism of Transport	Symbolic Description	Known Number after Transport	Equation	Unknown Number (Interpretation)
Bat	Time bat enters water	River	$3t$	Distance traveled = 15 miles	$15 = 3t$	5 hours before
Person	Time person leaves a spot	Automobile	$50t$	Distance traveled = 200 miles	$200 = 50t$	4 hours before
Number	Number chosen	Set of numerical instructions	$15x + 20$	125	$15x + 20 = 125$	7

Note that the person in the automobile is a new problem corresponding to a car traveling at a constant speed of 50 mph over a total distance of 200 miles.

It is common to capture the relationship between distance, speed, and time as the well-known formula already mentioned several times in this chapter: $d = st$. It is less common to represent as a formula the relationship between the unknown number chosen by the student and the calculated number, but we can do so if desired. The relationship is given by $15x + 20$ = calculated number. This can be abbreviated to $15x + 20 = C$ (or equivalently $C = 15x + 20$) with the situation in the table corresponding to $C = 125$. If the calculated value obtained by the class had been 170 instead, then the teacher could have placed $C = 170$ into the equation to obtain $170 = 15x + 20$. Solving this would reveal the number chosen by the student to be 10. The formula for C contains the legs to handle any such situation where the instructions are to multiply the unknown whole number by 15 and then to add 20.

If desired, we could interpret the bat's travel in the river as obeying a set of algebraic instructions:

	Steps of the Procedure	Algebraic Interpretation at Each Step
Step 1	Pick a number whose value is unknown	t
Step 2	River multiplies this number by 3	$3t$
Step 3	Distance value obtained is 15	$15 = 3t$
Step 4	Solve the equation	$5 = t$
Step 5	Time of travel in the river is 5 hours	

On the flip side, we could metaphorically think of the unknown number chosen by the student as being transported—via the teacher's instructions—to the number 125.

Though we risk pushing the metaphors a bit too far here, these viewpoints can still, at times, be leveraged for conceptual gain. Let's now see if we can use some of these ideas to build a framework in which to better understand an important physical phenomenon.

RADIOACTIVITY: REGULARITY OUT OF SPONTANEITY

Radioactivity was discovered in early 1896 by French physicist Antoine Henri Becquerel. He accidentally discovered it while investigating another phenomenon called phosphorescence.

Phosphorescent substances are materials that absorb light of a certain energy and slowly re-emit it, usually as light of a different energy. Many items that glow in the dark exhibit this type of behavior, where, for example, absorbed white light gets re-emitted over time as a soft green light that is most readily seen in the dark.

Becquerel was investigating uranium for this type of behavior when he realized that it wasn't necessary to first shine light on the element. Uranium that was kept in the dark for days—long after all phosphorescent effects would have ceased—was still emitting something, suggesting that the emissions were caused by some source internal to uranium itself and not due to some external agitation.

Very soon afterward, it was found that other elements possessed this property as well. This would turn into one of the most sensational and consequential discoveries in the history of science, for in time it led to the realization that the internal structures of the elements were much more sophisticated and energetic than originally thought and that they existed in different varieties called isotopes.

The newly discovered property was eventually christened with the name *radioactivity* by the renowned physicists Marie and Pierre Curie, who along with Becquerel shared the 1903 Nobel Prize in Physics for their groundbreaking discoveries.

Merriam Webster defines radioactivity as "the property possessed by some elements (such as uranium) or isotopes (such as carbon-14) of spontaneously emitting energetic particles (such as electrons or alpha particles) by the disintegration of their atomic nuclei."[3] Depending on the nature of the isotope, these emissions, called radiation, can be quite complex. The three most common and naturally occurring types of radiation are alpha particles (nuclei of helium), beta particles (electrons), and a form of high-energy light known as gamma rays (gamma photons). See the next three images (not drawn to scale):

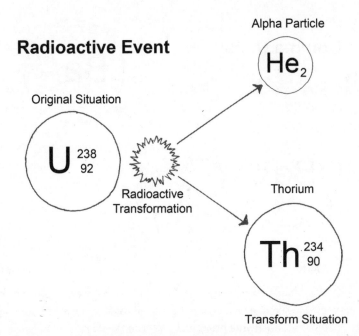

Uranium-238 atom changing identity (disintegrating) by emitting an alpha particle (helium nucleus) and turning into an isotope of the element thorium.
[Artwork provided by William Hatch]

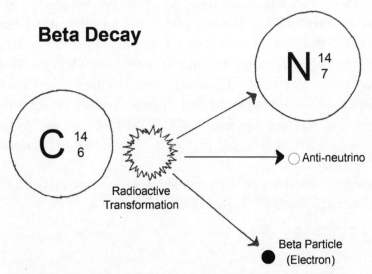

Carbon-14 atom changing identity (disintegrating) by emitting a beta particle (electron), anti-neutrino, and turning into an isotope of nitrogen.
[Artwork provided by William Hatch]

Gamma Decay

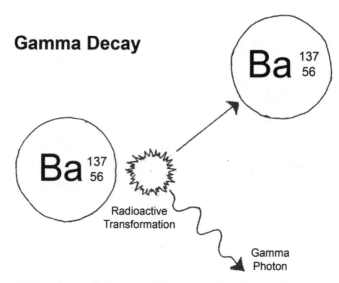

Barium-137 emitting a high-energy light particle (an electrically neutral gamma photon) and moving to a lower energy state but retaining its atomic identity.
[Artwork provided by William Hatch]

Behaviors such as these mean that materials containing these radioactive atoms will have the number of such atoms reduced over time, as the affected atomic nuclei undergo emissions that transform their atomic identity into another element. For radioactive atoms such as Barium-137 that don't change their identity, only their internal energy, the number of such atoms in a higher energy state will be the thing that decreases as a result of the emissions. In either case, the effects of these changes of state—changes in mass, weight, or energy—can be described numerically, meaning that the mass, weight, or energy of the radioactive substances vary over time, resulting in numerical symphonies.

A numerical ensemble for a hypothetical radioactive material (Imaginary Material-1 or IM-1) is given here:

Amount Present (Ounces)	Time (Years)
512	0
256	1
128	2
64	3
32	4
16	5

Numerical ensemble for decay of radioactive material (IM-1)

From an initial amount of 512 ounces, the amount of IM-1 decreased by half to 256 ounces in one year. In fact, notice that each year the amount of IM-1 present decreases or decays by half from the previous year. All radioactive isotopes decay in a similar fashion, which can be described by a characteristic number that tells, on average, how long it takes for a pure quantity of such substances to decrease by half. This length of time is called the *half-life* of the material and was one of several discoveries made during the earliest years of the twentieth century by physicist Ernest Rutherford and chemist Frederick Soddy. Both men were eventually awarded Nobel Prizes in Chemistry for this and other work on radioactive substances: Rutherford in 1908 and Soddy in 1921. (Rutherford had the singular good fortune to have his most important contribution to science come after already receiving a Nobel Prize— the identification of the atomic nucleus in 1911.)

Thus, IM-1 has a half-life of one year. Other isotopes of various elements have different half-lives, which can range from times far less than a billionth of a second to times far in excess of a billion years.

REGULARITY OF THE RIVER IS LINEAR

Regularity implies predictability. For example, in the following table, the river through its 3-mph current is able to connect the time passed with the total distance the bat has traveled, in a way that is measurable and symbolically described. This means that we can use either the time passed to predict the distance traveled or vice versa. Analogously, radioactive decay through its half-life regularity is able to connect the time passed with the amount of IM-1 left, allowing us to use either time as a predictor of the amount left or, more pertinently, the amount of substance left as a predictor of the time passed.

Distance (Miles)	Time (Hours)
0	0 (8:00 AM)
3	1 (9:00 AM)
6	2 (10:00 AM)
9	3 (11:00 AM)
12	4 (12:00 PM)
15	5 (1:00 PM)

Numerical ensemble for river travel

As discussed in Chapter 8, tables and algebraic expressions can be thought of as two ways of capturing or representing variable behavior. In the case of the table for river travel, we already know the expression that captures the ensemble: $d = 3t$. You can check this by letting $t = 1, 2, 3, 4, 5$ hours and observing that the distances match in the table. The table represents snapshots at the five listed times of the general varying behavior.

Once we have the formula, we can, of course, more readily solve for values not listed in the table such as the distance traveled by the bat after 2.5 hours or 7.2 hours, if not fished out by the rafter. These respectively yield $d = 3(2.5) = 7.5$ miles and $d = 3(7.2) = 21.6$ miles.

RADIOACTIVE REGULARITY IS EXPONENTIAL

What is the formula describing the situation involving the radioactive material IM-1? Even if we don't know it, we can still use the table to make estimates for the values we don't have. For example, if we want to know how much of the material is left after 2.5 years—halfway between the shaded entries for 2 and 3 years—we could guess 96 ounces, the amount halfway between 128 and 64 ounces.

The latter two weights correspond to the amounts of radioactive material left after the snapshots taken at 2 and 3 years, respectively. Our guess here is incorrect, however, as the actual value left after 2.5 years would be very close to 90.5 ounces.

Nevertheless, the table still arms us much better than being without it by allowing us to do an approximate type of algebra—sometimes called *linear interpolation*—that yields bounded guesstimates that aren't wildly outlandish.[4]

The actual formula for describing the numerical ensemble generated by the decay of IM-1 in the table is given by amount of IM-1 $= 512\left(\dfrac{1}{2}\right)^{t}$.

In the formula, "amount of IM-1" plays a role similar to the one distance did earlier in that it independently exists on its own, is measurable, and can be viewed as being simply related to the information on the right-hand side of the equation. In other words, the communication between both sides of the equation is two-way. We will abbreviate it to A, which gives $A = 512\left(\dfrac{1}{2}\right)^{t}$. Let's observe that this generates values in the table by considering $t = 3$ years. Replacing t by 3 in the formula yields

$$A = 512\left(\frac{1}{2}\right)^{3} = 512\left(\frac{1}{2}\right)\left(\frac{1}{2}\right)\left(\frac{1}{2}\right) = 512\left(\frac{1}{8}\right)$$

$$= \left(\frac{512}{1}\right)\left(\frac{1}{8}\right) = \frac{512}{8} = 64 \text{ ounces.}$$

This matches the amount given in the table for 3 years. The other amounts can be similarly calculated.[5]

Now consider a question similar to those we asked about finding the time of travel of the bat. How long would it take for the initial 512-ounce radioactive sample to decay, decrease, or "travel" to a point in time where there are only 11 ounces of IM-1 left?

For this problem, it would mean that $A = 11$ and t would be unknown. The formula for IM-1 becomes $11 = 512\left(\dfrac{1}{2}\right)^{t}$. Mathematically, we now want to solve this equation for t.

This equation is different from any that we have solved in this book so far because the unknown is in an exponent. Such equations where the variable is in the exponent are known as *exponential equations*.

The techniques developed in Chapter 3 will completely fail here. Those methods work if the equation involves $\frac{1}{2}$ times t, which is written as $\left(\frac{1}{2}\right)t$. They do not work for an exponential equation, which here involves $\frac{1}{2}$ raised to the t power, which is written as $\left(\frac{1}{2}\right)^t$.

Remember that for $t = 5$, $\frac{1}{2}$ times t becomes

$$\left(\frac{1}{2}\right)5 = \left(\frac{1}{2}\right)\left(\frac{5}{1}\right) = \frac{1 \cdot 5}{2 \cdot 1} = \frac{5}{2}.$$

Conversely, $\left(\frac{1}{2}\right)^t$ for $t = 5$ becomes

$$\left(\frac{1}{2}\right)^5 = \left(\frac{1}{2}\right)\left(\frac{1}{2}\right)\left(\frac{1}{2}\right)\left(\frac{1}{2}\right)\left(\frac{1}{2}\right) = \frac{1}{32}.$$

These values are not equal; in fact, $\frac{5}{2}$ is 80 times larger than $\frac{1}{32}$.

Students generally learn the maneuvers for solving exponential equations in an intermediate or college algebra class (Algebra 2 or above in high school), and the use of logarithms is generally involved. The result of employing these maneuvers—not shown—to solve the equation $11 = 512\left(\frac{1}{2}\right)^t$ is a t value of approximately 5.54 years.

Because showing the steps involved in solving these types of equations will not be attempted, our conceptual flight continues in our ensuing discussions of this situation. The broad ideas for using exponential equations to discover unknown information—in this case solving for t—are the same as for the equations we have solved in this book, they just differ in the details. Let's see what points of contact we can still see from taking only an overview of this material.

BIG ALGEBRA FROM THE AIR

Though we won't be discussing the techniques for solving exponential equations here, we can still apply some of what we have discussed throughout the book to gain insight. This is particularly true in trying to classify two of the major types of scenarios involving radioactive substances.

One involves the amount of material we initially start with—which can vary from scenario to scenario—and the second involves the type of radioactive substance used—which also can vary from scenario to scenario. Both of these can be characterized and tuned, respectively, by a specific parameter.

INITIAL AMOUNT

Let's begin by considering the initial amount of a substance. For now, let's stay with IM-1. Originally, we started with 512 ounces of material, which then decreased following a specific decay pattern. If we had started instead with 816 ounces of IM-1, then another pattern—with different amounts for a given year—would ensue:

Amount Present (Ounces)	Time (Years)
816	0
408	1
204	2
102	3
51	4
25.5	5

Second numerical ensemble for decay of radioactive material IM-1

Note that here the amounts also decrease to half the amount of the previous year.

The formula for the numerical symphony generated in this scenario is given by $A = 816\left(\dfrac{1}{2}\right)^t$.[6] So, for a given scenario involving IM-1, the initial amount is a constant, but it can vary from scenario to scenario

(512 ounces in Scenario 1 and 816 ounces in Scenario 2). This means that the initial amount of material present behaves like a parameter.

There are many ways in which to choose a letter to represent this parameter, but because it describes an amount too, it can be advantageous to be suggestive of this in our choice of symbol. However, we have to also be mindful to distinguish this from the A that we are already using in the formula.

A standard way to represent this initial amount is one that designates it as the value of the amount A when time equals 0 (at the start of the process). This could be done as "A at $t = 0$," which we can abbreviate to $A0$. However, it is common practice to set the 0 as a subscript to further distinguish it (for instance, to not be confused with multiplication), which gives A_0.

Inserting this into our formula for IM-1 yields the big algebra formula for IM-1:

$$A = A_0 \left(\frac{1}{2}\right)^t.$$

For the first IM-1 table, we would let $A_0 = 512$, and for the second one, we would let $A_0 = 816$.

If we started instead with an initial value of 27 ounces of IM-1, we would have $A_0 = 27$, which when substituted in the formula gives $A = 27\left(\frac{1}{2}\right)^t$. Thus, the general formula has the legs to describe any initial amount scenario for IM-1. We simply set A_0 equal to that initial value, and then the formula will take over predicting how it decays through time after that.

TYPE OF RADIOACTIVE MATERIAL

Let's now consider a different radioactive isotope, IM-3, that has a half-life of three years. If we initially have 816 ounces of this isotope, then the decay pattern will look like the following:

Amount Present (Ounces)	Time (Years)
816	0
408	3
204	6
102	9
51	12
25.5	15

Numerical ensemble for decay of radioactive material IM-3

Note that the amount of IM-3 has decreased by half to 408 ounces in three years (not one year, as with IM-1).

The formula that worked for IM-1, $A = 816\left(\dfrac{1}{2}\right)^{t}$, will not accurately predict the slower decay pattern for IM-3, but something closely related will, namely $A = 816\left(\dfrac{1}{2}\right)^{\frac{t}{3}}$. Check that when $t = 6$ years, the latter formula will yield the value listed in the table:

$$A = 816\left(\frac{1}{2}\right)^{\frac{6}{3}} = 816\left(\frac{1}{2}\right)^{2} = 816\left(\frac{1}{2}\right)\left(\frac{1}{2}\right) = 816\left(\frac{1}{4}\right) = \frac{816}{4} = 204\,\text{ounces.}$$

You can check that this will work for all of the other times as well: $t = 0, 3, 9, 12, 15, \ldots$[7]

This second type of scenario depends on what isotope we have and will change from one isotope to another depending on the half-life of the isotope in question. This means that for a given isotope the half-life is constant, indicating that it is parameter-like in its behavior, too. To represent this parameter, we need to introduce a new letter. We will opt again in favor of being descriptive, designating h for half-life.

For an initial amount of 816 ounces, this will yield the formula $A = 816\left(\dfrac{1}{2}\right)^{\frac{t}{h}}$. For $h = 1$ as in IM-1, we will have the formula $A = 816\left(\dfrac{1}{2}\right)^{\frac{t}{1}} = 816\left(\dfrac{1}{2}\right)^{t}$, whereas for $h = 3$ as in IM-3, we will have

$A = 816\left(\dfrac{1}{2}\right)^{\frac{t}{3}}$. For a third isotope IM-1205, with a half-life of 1205

years, we would have $A = 816\left(\dfrac{1}{2}\right)^{\frac{t}{1205}}$ (as $h = 1205$). Thus, after 1205 years

have passed ($t = 1205$), this formula would predict

$$A = 816\left(\dfrac{1}{2}\right)^{\frac{1205}{1205}} = 816\left(\dfrac{1}{2}\right)^{1} = 816\left(\dfrac{1}{2}\right) = \left(\dfrac{816}{1}\right)\left(\dfrac{1}{2}\right) = \dfrac{816}{2} = 408\,\text{ounces}$$

(which is half of 816 ounces after 1205 years).

Combining this information with that on the initial amount in the previous subsection leads to an even more general formula that has the reach to describe any initial amount (A_0) of a radioactive material with any half-life (h) as

$$A = A_0\left(\dfrac{1}{2}\right)^{\frac{t}{h}}.$$

For 320 ounces ($A_0 = 320$) of the real and dangerous isotope strontium-90, which has a half-life of approximately 28.8 years ($h = 28.8$), this more general formula would become $A = 320\left(\dfrac{1}{2}\right)^{\frac{t}{28.8}}$. For 1237 ounces of the most common isotope of uranium, uranium-238, which has a half-life of approximately 4.5 billion years, the formula would become $A = 1237\left(\dfrac{1}{2}\right)^{\frac{t}{4,500,000,000}}$. And so on.

RADIOCARBON DATING

Because the regularity of radioactive decay connects the amount of material left to the time passed—as discussed earlier—the mechanism can serve as a trailblazing approach to measuring the passage of time. This clocklike regularity of radioactive materials is based on the steady predictable diminishing of the amount of radioactive substance present via half-life periods. This diminishing can be measured in a number of ways, including diminishing mass, weight, or energy of the radioactive material over time (illustrated in the previous two sections), diminishing intensity of radioactive emissions over time, or diminishing ratio of radioactive material to nonradioactive material present over time.

If it helps, we can think of radioactive behavior as possessing a counting type of regularity that eventually zeros out as the number of half-life cycles (HLC) for the material increases: 1 HLC (50% left), 2 HLC (25% left), 3 HLC (12.5% left), 4 HLC (6.25% left), and so on.

In the 1940s and 1950s, powerful techniques that took advantage of these regularities were introduced and developed by chemist Willard Libby and his coworkers. He received the 1960 Nobel Prize in Chemistry for these efforts.

To see how it works, let's consider a box that contains 92 ounces of the nonradioactive stable isotope IM (which doesn't decay) plus 8 ounces of the radioactive isotope IM-5730 (which decays by emitting an electron and becoming a new element) with a half-life of 5730 years. These two isotopes combine for a total of 100 ounces of substance in the box initially.

For the sake of simplicity, let's assume that after an atom of IM-5730 undergoes radioactive decay, the new element(s) produced along with anything emitted in the process are immediately pumped out of the box. Furthermore, we assume that for every IM-5730 atom that disappears via this radioactive decay, we can immediately replace it by pumping in a new one. This means that the ratios in the box stay the same—or are in equilibrium—as long as the pump stays on: namely, 92 ounces of the material in the box is stable IM and 8 ounces is radioactive IM-5730.

However, if we permanently turn the input pump off, then the amount of the radioactive IM-5730 will decrease over time, and as a consequence so will the corresponding percentage of IM-5730 versus the original 8 ounces in the box. We illustrate this in the following table:

Amount Left (Ounces)	Percentage of Initial Amount Left	Time after Pump Stops (Years)	Half-Life Cycles
8	100%	0	0
4	50%	5730	1
2	25%	11,460	2
1	12.5%	17,190	3
0.5	6.25%	22,920	4
0.25	3.125%	28,650	5

Numerical ensemble for amount of IM-5730 left after input pump stops

Note that the initial amount of IM-5730 decreased by half to 4 ounces after 5730 years; the initial amount decreased by one-fourth to 2 ounces in (5730)(2) = 11,460 years.

If someone came upon such a box that had been around for a long time and found that the amount of IM-5730 left was only 12.5% of the normal amount expected, then they could deduce—through knowledge of how radioactive mechanisms work—that the input pump turned off 3 HLC ago, or 17,190 years before. Or, if we invoke the metaphor, it would take 17,190 years for the initial amount of 8 ounces to be transported—via radioactive mechanisms—to the current amount of 1 ounce.

If someone found another such box with 1.64 ounces left (or 20.5% of the normal amount) and wanted to figure out when the input pump for it stopped, they could tune the parameters in the half-life equation, $A = A_0 \left(\dfrac{1}{2} \right)^{\frac{t}{h}}$, by setting A_0 = 8 and h = 5730, which would yield

$A = 8 \left(\dfrac{1}{2} \right)^{\frac{t}{5730}}$. For the specific situation of 1.64 ounces left, we would

substitute in A = 1.64, which gives $1.64 = 8 \left(\dfrac{1}{2} \right)^{\frac{t}{5730}}$. Then we would have

to solve this exponential equation for t. As we are not solving exponential equations in this book, we will simply give the correct value of t, which here would be approximately 13,100 years. This would mean that the pump turned off around 13,100 years prior to the discovery of the box.

As a check, you can see that 1.64 ounces (20.5%) lies between the shaded rows for 1 ounce (12.5%) and 2 ounces (25%) in the table for IM-5730, meaning that 13,100 years should fall between the predicted times for those two values in the table.

This is in broad scope how the technique, developed by Libby and his associates, called *radiocarbon dating* works. The human body consists of approximately 18% carbon, and this carbon comes in three varieties: two stable isotopes (carbon-12 and carbon-13) and one radioactive isotope (carbon-14, with a half-life of about 5730 years).[8] The ratio of carbon-14 to the stable carbon is extremely small but still measurable. As long as a body is alive, it replenishes through life processes the carbon-14 that decays away, keeping the ratio the same. However, when the body dies, the carbon-14 is no longer replenished—the input pump permanently turns off—and the ratio changes according to a half-life of 5730 years.

As the table for IM-5730 shows, the amount of material eventually becomes so small that there is a limit to what can be detected experimentally. For carbon-14, the detectable amounts are evidently good up to about 50,000 years, which corresponds to slightly less than 1% of the amount of carbon-14 present when the body was alive.[9]

Other techniques, such as mass spectrometry, can be used for dating organic materials beyond the limit of the method just described. It is also possible to use still additional isotopes to date some nonorganic objects such as rocks. The procedures differ according to the isotope used, but the general idea of measuring the percentage of radioactive material left versus an initial amount of material is the same. The method of using radioactive materials as time markers is called *radioisotope* or *radiometric dating*. It goes without saying that it is a method whose value to science and human understanding is monumental.

TABULAR ALGEBRA

The examples in the previous sections show that it is possible to use tables to make not-too-outlandish estimates of unknown information. We did this as a commonsense verification of the 13,100 year value on the time it would take 8 ounces of IM-5730 to decay to 1.64 ounces.

However, there is nothing to prevent us from using tables on their own, and systematically, as another way to gain insight about unknown numerical information from an array of known values. Such an approach would have the advantage that it doesn't require intimate knowledge of how to algebraically solve exponential equations to get an approximation.

We can think of techniques used with tables this way as being a type of "tabular algebra," analogous in spirit to the way we take an ensemble of numbers generated by an algebraic expression (say $6t - 30$) and find out through algebraic maneuvers what value of t makes $6t - 30$ become, say, 180 (by solving $6t - 30 = 180$): in other words, systematically finding or approximating unknown values from known ones.

The techniques for using algebraic expressions give an exact solution to the algebraic equations discussed so far, whereas the techniques involving tables generally only give approximate solutions for exponential equations. A way to make the tabular technique yield even better estimates is to increase the resolution of the table, by filling in more of the gaps through expanding the original table into a larger, more detailed table. For example, in the case of IM-5730, someone with the right expertise could calculate the time it should take the 8 ounces to decay to amounts that additionally include the in-between values for each of the six previous table entries (8, 4, 2, 1, 0.5, and 0.25 ounces). Some in-between values in this case would be 6, 3, 1.5, 0.75, 0.375, and 0.125 ounces. Doing this will yield new information in the shaded cells:

Amount Left (Ounces)	Percentage of Initial Amount Left	Time after Pump Stops (Years)
8	100%	0
6	75%	2378.2
4	50%	5730
3	37.5%	8108.2
2	25%	11,460
1.5	18.75%	13,838.2
1	12.5%	17,190
0.75	9.375%	19,568.2

0.5	6.25%	22,920
0.375	4.6875%	25,298.2
0.25	3.125%	28,650
0.125	1.5625%	34,380

Expanded numerical ensemble for amount of IM-5730 left
after input pump stops

Interpolation using this more detailed table gives an estimate of 12,649 years for the time it would take 8 ounces of IM-5730 to decay to 1.64 ounces. This is closer to the true value of 13,100 years than is the estimate obtained from using the earlier, less detailed table—with only six entries—which interpolates to 14,325 years (see endnote for a calculation of both estimates).[10]

If desired, more values can be included to build even larger tables that will yield even better approximations. If enough values are filled in, then these higher-resolution tables can ultimately, for practical uses, give almost as accurate a result as the formulas themselves.

Increasing the resolution of our tables by adding more rows of information is quite analogous to how adding more pixels—or data values—to a digital photograph increases the resolution of the image—improving its ability to better replicate the subject.

Remember that the computed values here all come from the formula $A = 8\left(\dfrac{1}{2}\right)^{\frac{t}{5730}}$. Moreover, it is the percentages that are really the key, as these will hold regardless of the initial amount; that is, if we instead initially had 500 ounces of IM-5730, it would still take nearly 23,000 years for this new amount to decay to 6.25% of the amount we started with (31.25 ounces of IM-5730 in this case).

Using tables to obtain the information is probably the way that most people would deal with this particular situation (like they use tax rate tables to compute their income tax), but it is still more helpful than not knowing that algebraic expressions serve as a powerful force in underwriting the technique. For those who even mildly better understand these expressions, especially regarding scenario variables or parameters, it may give them the ability to more critically examine the situations presented to them.

This is especially important for decision makers who must deal with information from tables that describe social situations, demographics, economics, and so forth, where the laws are not so well defined and more assumptions—including the very formulas themselves in many cases—have to be made.

OTHER RENDITIONS OF EXPONENTIAL SONGS

The essence of the way in which radioactive materials regularly decrease in amount is exhibited in a wide assortment of phenomena. Just as "The Star-Spangled Banner" connects its different renditions—varying in tone, quality, and pace—through its lyrics, so too again does algebra allow us to connect these widely different phenomena through the agency of its expressions, of exponential type in this case.

Before diving in, let's remember that in its simplest forms—using whole numbers—an exponent means repeated multiplication. For instance, a shorthand way to write the multiplication

$$\underbrace{2\times2\times2\times2\times2\times2}_{\text{six 2s multiplied}}$$

is through the shorthand notation 2^6. Here, the 2 in the base means the number that we are multiplying, and the 6 in the exponent tells us how many 2s are involved in the multiplication.

Going back to the half-life of IM-1, we have the situation that each amount is one-half of the amount for the preceding period (preceding year in this case). This leads to the repeated multiplication of $\frac{1}{2}$, which results in powers of $\frac{1}{2}$ as demonstrated in the following table.

A_t (Amount in Year t)	Time (Years)	Relation to Previous Year	In Terms of Initial Amount (A_0)	In Exponential Form
A_0	0	Initial amount (year 0 amount)	A_0	A_0
A_1	1	$\left(\frac{1}{2}\right)$ (year 0 amount)	$A_1 = \frac{1}{2}A_0$	$A_1 = \frac{1}{2}A_0$

A_2	2	$\left(\dfrac{1}{2}\right)$ (year 1 amount)	$A_2 = \dfrac{1}{2}A_1 = \dfrac{1}{2}\left(\dfrac{1}{2}A_0\right)$	$A_2 = \left(\dfrac{1}{2}\right)^2 A_0$
A_3	3	$\left(\dfrac{1}{2}\right)$ (year 2 amount)	$A_3 = \dfrac{1}{2}A_2 = \dfrac{1}{2}\left(\dfrac{1}{2}\right)^2 A_0$	$A_3 = \left(\dfrac{1}{2}\right)^3 A_0$
A_4	4	$\left(\dfrac{1}{2}\right)$ (year 3 amount)	$A_4 = \dfrac{1}{2}A_3 = \dfrac{1}{2}\left(\dfrac{1}{2}\right)^3 A_0$	$A_4 = \left(\dfrac{1}{2}\right)^4 A_0$
A_5	5	$\left(\dfrac{1}{2}\right)$ (year 4 amount)	$A_5 = \dfrac{1}{2}A_4 = \dfrac{1}{2}\left(\dfrac{1}{2}\right)^4 A_0$	$A_5 = \left(\dfrac{1}{2}\right)^5 A_0$

This yields the general formula for t years: $A_t = \left(\dfrac{1}{2}\right)^t A_0$ or equivalently

$A_t = A_0\left(\dfrac{1}{2}\right)^t$. We usually drop the subscript t and write $A = \left(\dfrac{1}{2}\right)^t A_0$ or

$$\text{equivalently } A = A_0\left(\dfrac{1}{2}\right)^t.$$

How formula is obtained for IM-1

Though the $\dfrac{1}{2}$ is crucial to the way that radioactive materials decay, it is one of a myriad of possibilities in terms of the mathematics alone. If we have a situation such that each amount is triple the previous amount in the year before, then $\dfrac{1}{2}$ simply gets replaced by 3 and we will get the general formula for t years as $A = (3)^t A_0$ or equivalently $A = A_0(3)^t$.

Such situations can occur in the way money grows in certain interest-bearing accounts. For instance, if one invests in an account that earns 2% interest compounded annually, then this means that every year, 2% interest is applied to the amount from the previous year. Thus, the total amount in the account for, say, year 1 is the amount from year 0 plus 2% of the amount from year 0. In symbols, this becomes $A_1 = A_0 + 0.02A_0$, which simplifies to $A_1 = 1.02A_0$ (think $A_0 + 4A_0$ simplifying to $5A_0$). Reasoning similarly will give $A_2 = 1.02A_1$, $A_3 = 1.02A_2$, and so on.

Here, the multiplier every year is 1.02, which means that in the formula from the table, we simply can replace $\frac{1}{2}$ by 1.02 to get the formula for how much would be in the bank account after t years as $A = (1.02)^t A_0$ or equivalently $A = A_0(1.02)^t$.

If the interest rate instead were 4.5%, we would now have a multiplier each year of 1.045: for this case, $A_1 = A_0 + 0.045A_0$ simplifies to $1.045A_0$, $A_2 = A_1 + 0.045A_1$ simplifies to $1.045A_1$, and so on. The formula for how much would be in the bank account after t years would now be $A = (1.045)^t A_0$ or equivalently $A = A_0(1.045)^t$.

These two examples show that we have the interest rate varying from scenario to scenario but constant within a specific scenario, meaning that it behaves as a parameter. Introducing a scenario variable for it and opting for being descriptive again, we will call it r. We will assume that r is given in decimal form. Doing so means that we have the general formula $A = (1 + r)^t A_0$ or equivalently $A = A_0(1 + r)^t$. This general formula now tells us the amount we will have after t years if we invest an initial amount of A_0 dollars at an interest rate of r (given as a decimal). For the two situations described, we have $r = 0.02$ and $r = 0.045$, respectively.

Consider the scenario where we invest $30,000 into an account that earns 1% interest compounded annually, and we want to know how much we will have after 10 years. We simply use the general formula with the following scenario settings: $A_0 = 30{,}000$ and $r = 0.01$. This would give us the specific formula for this scenario as $A = 30000(1 + 0.01)^t$. The specific time that we are interested in for this scenario is 10 years. Substituting $t = 10$ into this formula will give us $A = 30000(1 + 0.01)^{10} = 30000(1.01)^{10} = \$33{,}138.66$.

Thus, the money would have grown by $3,138.66 over the 10-year period. We could use the expression for this scenario to find out values of the investment for other times as well, such as $t = 15$ years or $t = 20$ years.

Many other phenomena—population growth for a time, initial microorganism growth, and early spread of a disease or a rumor, for example—may grow or decrease in similarly regular patterns. And for these, we can also use exponential expressions to predict their behavior and then put them in tables if need be to aid in the sharing of this information.

A MOST REMARKABLE PROPERTY OF EXPONENTIAL BEHAVIOR

More times than not, when you find the general exponential formula describing radioactive decay—or many other types of exponential decay or growth—it will not be given in the form that we have elaborated on in this chapter, but rather in the form of something like $A = A_0 e^{kt}$. The base of this exponent is the number e, which is given by 2.718281828459045…

The number e is an irrational number (like π or $\sqrt{2}$) whose decimal expansion goes on forever without a repeating pattern of digits. This means that the amount could be approximated by rounding e off to 2.71828, making the expression become $A \approx A_0 (2.71828)^{kt}$. The symbol $\approx$ stands for "approximately equal to." However, most calculators nowadays have a key for giving better approximations of the number e as well as the exponential expression e^{kt}, which means that we can use these keys as opposed to writing out the less accurate $(2.71828)^{kt}$.

This, of course, differs from the earlier formula

$$A = A_0 \left(\frac{1}{2}\right)^{\frac{t}{h}}.$$

Though the formula with a base of e is certainly less natural in the conceptual sense, it is worth noting that there are still two parameters present, this time A_0 and k. The A_0 is the same as in our earlier discussions, but the k—though still related to the half-life of the radioactive isotope—is not the same as the half-life of the isotope h.

Why do mathematicians and scientists choose to use this strange-looking number e as the base of exponential expressions?

This occurs in part due to a few important factors. Firstly, it is possible to represent exponential formulas of any base by one exponential formula of a single base. And secondly, exponential formulas that involve the number e as a base more naturally capture key components related to phenomena that exhibit a continuous type of growth or decay (where they occur all of the time) versus exponents to other bases. Exponential formulas to base e also have extremely nice properties with respect to the primary operations in calculus—differentiation and integration—and all that follows from this. See Appendix 4 for a brief discussion on how an exponent of one base can mimic exponents with different bases.

CONCLUSION

We have covered a diverse trail of algebraic and conceptual terrain in this chapter. Owing to the big gaps in the particulars—regarding some of the maneuvers employed—the approach taken here has been likened to taking a flight over those topics.

Has anything been gained from doing this? More specifically, has one of the book's goals of providing fertile soil for the reader—to better coordinate, better appreciate, and possibly better manipulate—numerically varying phenomena been achieved?

As always, only readers can answer these deeply personal questions for themselves. In this conclusion, we will discuss only a few of the possible takeaways.

We started by focusing on the idea of mechanisms and how we can use our understanding of their effects, even if we don't fully understand their inner workings, to acquire new information. This in a sense then became a metaphor for the entire chapter, where the fundamental idea was to take numerically varying phenomena as encountered—and whose inner workings may not have been completely understood in some cases—and still be able to think about or organize them in some productive way: a way that hopefully aided in better understanding the phenomena conceptually. This involved getting down to the nitty-gritty details in some cases while only understanding the gist of the phenomena in others.

We now summarize a couple of the perspectives that were enlisted to aid in this endeavor.

THE MECHANISM TRANSPORT VIEWPOINT

This framework involved viewing an unknown number as being "transported" or "related" via some type of procedure or mechanism to another known number of interest. The aim was to find the unknown number, which in the cases here meant modeling the situation by an algebraic equation and then solving it.

We started out with a purely algebraic mechanism—the teacher's instructions—but saw afterward that other physical phenomena can be productively thought of in this way, too. In those cases, the mechanism itself wasn't a set of algebraic instructions but was of a nature that it could still be modeled by such instructions—presented as algebraic expressions—thus opening up their analysis to the many techniques available on the mathematical grid.

Going the reverse way, we discussed the prospect of using a particular physical situation as a metaphor to productively think about other physical situations and the algebra itself. For this, we chose the foundational situation to be an object traveling or being transported via the medium of a tangible river from an initial or starting point to a final point.

All of this allowed for two-way connections to be made between various types of phenomena, using one to assist in understanding and working with the other and vice versa. This transport viewpoint is particularly well suited to phenomena that continually decrease or increase in time—or in relation to some other variable—such as the amount of radioactive material present in time or the distance traveled in time, respectively.

The process of leveraging algebra—and math in general—to discover unknown information, though predating the European Renaissance by centuries, gained more expansive tools in the 1500s and 1600s and was looked upon afresh as a powerful new type of reasoning: a brand that allowed for the systematic and unassailable discovery of hidden knowledge. It was a type of reasoning that, when joined together with the emerging experimental sciences, became so seductive and commanding that it took the seventeenth- and eighteenth-century world by storm, ultimately leading to the almost total downfall of other forms of reasoning about the physical world.

This included certain aspects of scholasticism—the dominant form of scholarly logical thinking in Europe throughout the late medieval period—a form that ultimately became the favorite target of criticism by later, more investigation-inclined natural philosophers and scientists.[11]

THE PARAMETER VIEWPOINT

Parameters are certainly one of the more commonly under-recognized and underappreciated tools in the algebraic arsenal. Throughout this text, we have shined a spotlight on them to illustrate how useful they can be for better classifying, understanding, and working in an all-encompassing way with numerically varying phenomena.

Many people already use the essence of this idea as an organizational tool in some familiar settings. Consider the way that some of us deal with television programming, which changes throughout the day. Rather than attempting to understand head on the entire totality of all programs showing in a given day, many try to organize them in more convenient and accessible ways. Two of the most natural involve looking at the program scheduling by time of day or by particular station.

That is, we may want to know what programs are showing during a particular time of day. In this case, we fix the time and allow the channels to vary, treating each hour or half-hour as a distinct scenario to be considered and allowing the channels to vary in each scenario. In other cases, we may want to know what programs are showing on a particular channel throughout the day. In the latter, we fix the channel and allow the time to vary, treating each channel as a distinct scenario.

Either of these are useful vantage points by which to partition television programming into digestible and useful chunks without trying to encompass the totality of all that is showing on every channel at every time for the entire day. Other organizing possibilities, of course, include searching programs by fixing categories such as sports, history, news, and documentaries.

Organizing locations on Earth according to longitude and latitude can present another example of acquiring useful information by fixing one item while allowing another to vary. If we fix the latitude and allow the longitude (east or west location) to vary, we get imaginary circles that geographers call parallels (circles parallel to the equator). Each parallel describes the set of all the locations on Earth that are at the same north or south position of the equator (0° latitude). If instead we fix the longitude and allow the latitude (north or south location) to vary, we get imaginary half-circles that geographers call meridians. Each meridian describes the set of all locations that are at the same east or west location on Earth with respect to their degree separation from the prime meridian (0° longitude).

Using parameters in algebra give us the ability to intensify such classification schemes by making them more precise and operational. Once a set of parameters are identified, we can then inject them into algebraic expressions, equations, and tables to fix a particular scenario and then have this result interact productively with other types of variation within the scenario.

In the case of radioactive phenomena, the two classification settings that we looked at were the initial amount of a radioactive substance present and the half-life or rate of decay of a particular substance. By using parameters to describe them, we were able to incorporate these classifications into a single algebraic expression that had the legs to handle, in principle, the decay of any radioactive substance occurring in any initial amount. Here, we tuned both of those settings for a particular scenario and then viewed the time and the changing amount of the substance as the regular variables within each scenario.

Though we didn't discuss at any point how to actually solve the exponential equations involved in the description, this parameter description still allowed for a useful glimpse at the extremely complex phenomena of radioactivity in some of its most important and useful manifestations. That is, once the parameters are identified, we can still use them as classification tools even if we can't take complete advantage of all of the other mathematical tools available.

Unfortunately, though the idea of parameters permeates mathematics and science, we have already seen that their representation and treatment is anything but consistent in algebra. We discussed the Descartes protocol in Chapter 8, but as further discussion revealed, this protocol is often not followed outside of educational contexts due to the need to satisfy other requirements—and is not even always consistently followed within educational contexts.

Even what counts as a parameter is inconsistently applied. In some cases, there is a somewhat natural separation between what acts as a parameter and what acts as a regular variable. This is true in the case of radioactivity where the half-life and initial amount present naturally act as constants in a given scenario, whereas the time of decay and the amount of material remaining act as the natural variables. This is also the case in the distance, speed, and time situations we discussed, where the speed of travel was the natural parameter and the distance and time naturally acted as regular variables in a given scenario. Similarly, with break-even issues, the selling price, unit costs, and fixed costs act as natural parameters (constant in a given scenario), and the quantity of items sold acts naturally as the regular variable (varies within a scenario).

However, in other situations there may be no such obvious separation between what is a parameter and what is a regular variable. But this doesn't prevent us from still being able to impose a separation to aid in organizing the material. For instance, in the television programming situation, both station and time are in a sense equivalent—in that changing either usually yields a different program on the TV set—yet choosing to fix one or the other still can turn out to be very useful. This is also true for the longitude-latitude situation.

Irregularity in usage is something that occurs and causes confusion in many contexts. For instance, irregularity in the spelling and use of English words can often make learning to write in the language difficult. In a similar fashion, the irregularity in the alphabetic representation of parameters can make learning the conceptually rich and sometimes difficult subject of algebra even more difficult by presenting students with an alphabet soup of characters that are easy to get confused by.

The hope here is that, by continually acknowledging their existence and incorporating them into expressions and equations, parameters will perhaps seem a little less forbidding. Here is a list of some of the situations that we have discussed (or will discuss) in this book involving parameters:

Algebraic Object	Parameters: Scenario Variables (Constant within a Scenario but Change from Scenario to Scenario)	Regular Variables (Change within a Scenario)	Chapter
$Px = Cx + F$	P = price per item, C = cost per item, F = fixed costs	x = number of items sold	4
$100x + (2013 + z) - y$	z = years past 2013	x = number of days a week like to eat out, y = year of birth (for different people)	2
$ax^2 + bx + c = 0$	a = coefficient value of square variation, b = coefficient value of linear variation, c = constant coefficient value	x = variable or unknown number	Appendix 1
Course Average $= ax + by + cz$	a = homework contribution to final grade, b = test contribution to final grade, c = final exam contribution to final grade	x = homework average, y = test average, z = final exam score	8
GPA $= a_1x_1 + a_2x_2 + a_3x_3 + a_4x_4 + a_5x_5$	a_1 = point value for A grade, a_2 = point value for B grade, a_3 = point value for C grade, a_4 = point value for D grade, a_5 = point value for F grade	x_1 = % total credits at A, x_2 = % total credits at B, x_3 = % total credits at C, x_4 = % total credits at D, x_5 = % total credits at F	8
$A = A_0\left(\dfrac{1}{2}\right)^{\frac{t}{h}}$	A_0 = initial amount of material, h = half-life of material	t = time, A = remaining amount	9
$d = st$	s = speed of object	d = distance of travel, t = time of travel	9
$x^n + y^n = z^n$	n = positive integer power	x = first unknown, y = second unknown, z = third unknown	10

The idea of taking varying phenomena and thinking about them globally by slicing them up into "variation by scenario" pieces and "variation within a scenario" pieces is something that one should be aware of and on the lookout for in situations amenable to mathematical treatment.

This also includes circumstances when variation is displayed to us in tabular form. In many cases, where multiple tables or spreadsheets are used, each individual table or sheet may itself correspond to a fixed set of parameters—sometimes unrecognized as such—whereas variations within the table or sheet may correspond to the regular variables. Each of the tables we discussed on radioactivity can be so specified by fixing the parameters A_0 and h, with the time t varying within each sheet. Though these could all be generated from the formula $A = A_0 \left(\dfrac{1}{2}\right)^{\frac{t}{h}}$ with appropriate settings for A_0 and h, parameters for tables can still exist even when a formula for the behavior isn't known.

In the next chapter, we will take a look at algebra from a slightly different vantage point by examining some of the benefits that may accrue to mathematicians and others from initiating investigations out of sheer curiosity and prompt of imagination. The discussion will touch upon mathematical goings-on over a span of nearly 4000 years, briefly visiting ancient Mesopotamia, ancient Greece, Roman times, and the early scientific era, as well as the twentieth and twenty-first centuries.

Motions of the Imagination

Is mathematical analysis then...only a vain play of the mind?...Far from it; without this language most of the intimate analogies of things would have remained forever unknown to us; and we should forever have been ignorant of the internal harmony of the world...

—Henri Poincaré (1854–1912), "The Value of Science," *The Popular Science Monthly*

10

Algebraic Flights: Indeterminacy and Curiosity

Scientific subjects do not progress necessarily on the lines of direct usefulness. Very many of the applications of the theories of pure mathematics have come many years, sometimes centuries, after the actual discoveries themselves. The weapons were at hand, but the men were not ready to use them.

—Andrew Russell Forsyth (1858–1942), *Discussion on the Teaching of Mathematics Which Took Place on September 14th at a Joint Meeting of Two Sections*

To divide a cube into two other cubes, a fourth power, or in general any power whatever into two powers of the same denomination above the second is impossible, and I have assuredly found an admirable proof of this, but the margin is too narrow to contain it.

—Pierre de Fermat (1601–1665), written in the margin of his copy of Diophantus' *Arithmetica*

We next pivot our attention to a recurring and fascinating property that forms an important layer in the historical tapestry of algebra. We set the stage by returning to computing course averages from Chapter 8 and consider a fourth scenario with the following parameter settings:

Scenario	Weights/Contributions	Parameter Values	Course Average $= ax + by + cz$
4	20% from homework 60% from tests 20% from final exam	$a = 20\%$ $b = 60\%$ $c = 20\%$	$0.20x + 0.60y + 0.20z$

If one of the students has an overall course average of exactly 80% (with no rounding off), can we determine exactly how the student's scores divvy among the three categories of assignments; that is, can we say with absolute certainty what their average scores are on their homework, tests, and final exam, respectively?

Put another way, is there only one way in which a combination of the three average scores can unite to give an overall course average of exactly 80%?

Clearly, if the student had a straight 80% average across the board on every one of their homework assignments, as well as on all of their tests and their final, this would certainly combine to yield an 80% overall course average. This can be quickly verified: For $x = 80$, $y = 80$, and $z = 80$, we have

$$\text{course average} = 0.20(80) + 0.60(80) + 0.20(80) = 16 + 48 + 16 = 80.$$

There are other possibilities, however. A few of these are indicated here:

Averages	Variable Settings	Course Average $= 0.20x + 0.60y + 0.20z$	Course Average
Homework average = 70 Test average = 80 Final exam score = 90	$x = 70$ $y = 80$ $z = 90$	$0.20(70) + 0.60(80) + 0.20(90) =$ $14 + 48 + 18 = 80$	80
Homework average = 60 Test average = 80 Final exam score = 100	$x = 60$ $y = 80$ $z = 100$	$0.20(60) + 0.60(80) + 0.20(100) =$ $12 + 48 + 20 = 80$	80
Homework average = 95 Test average = 70 Final exam score = 95	$x = 95$ $y = 70$ $z = 95$	$0.20(95) + 0.60(70) + 0.20(95) =$ $19 + 42 + 19 = 80$	80
Homework average = 80 Test average = 90 Final exam score = 50	$x = 80$ $y = 90$ $z = 50$	$0.20(80) + 0.60(90) + 0.20(50) =$ $16 + 54 + 10 = 80$	80
Homework average = 0 Test average = 100 Final exam score = 100	$x = 0$ $y = 100$ $z = 100$	$0.20(0) + 0.60(100) + 0.20(100) =$ $0 + 60 + 20 = 80$	80

Thus, the answer to the question of whether or not we can determine with certitude what the average scores for the student are respectively on homework, tests, and the final exam—if we only know the overall average is exactly an 80—is negative; there actually turn out to be quite a lot of possibilities.

Mathematicians call situations where we can't pin down or determine a single set of values that yield a certain result—due to the existence of many possibilities—*indeterminate* situations. We have an indeterminate situation here due to the existence of several sets of different average scores for homework, tests, and the final exam that blend together to yield the same overall course average of exactly 80.

Translating the preceding situations into algebraic form shows that they are tantamount to being solutions to the equation $0.20x + 0.60y + 0.20z = 80$. The six different combinations of x, y, and z presented satisfy the equation, but there are many more. Because each of these combinations qualify to be called solutions, equations like this one are themselves called *indeterminate equations*.

We have yet another example of a class of equations that is unlike nearly all of the equations we have discussed throughout the book. The equations we have primarily focused on heretofore—the conditional ones and not identities—generally yielded a single, definitive solution.

One item that distinguishes our indeterminate equation from the earlier equations is that it has *three* variables. The other equations we solved had only *one* variable, or ultimately depended upon a single variable. The fact that the number of variables—three—exceeds the number of equations—one—is a significant feature. The extra variables here provide a flexibility and independence that allows many different combinations to work.

Like most things algebraic, indeterminacy is a general phenomenon that often crops up.

THE MAN FROM ALEXANDRIA

Way back sometime in the midst of the Roman era, the mathematician Diophantus did intriguing work with indeterminate situations. Hailing from Alexandria, Egypt, he is one of the shadowiest of the famous mathematicians from antiquity.

And the mystery starts with trying to figure out when he was even alive. The only definitive estimates pin his heyday to within a range of 500 years (150 BCE–350 CE). Another less definitive source places him as being alive around 250 CE, which is the date that many mathematical historians had accepted for a time and some still do.[1] His ethnicity is also uncertain.[2]

The amazing creativity of his work combined with its apparent isolation—from any similar known work so far—make it hard even today for mathematical historians to give a complete measure of his contribution to the algebraic narrative. The work in his famous book, *Arithmetica*, appears to have come out of nowhere, although it is viewed as being unlikely that all of it is due to him alone.[3]

Just as with Euclid's famous work, *The Elements*, and Al-Khwarizmi's famous text on algebra, the book is most likely a combination of earlier work by others as well as original work by the author himself, an extremely capable commentator and expert at mathematical logistics who added in his own explanations, outlook, organization, and perhaps problems and techniques, too.

In the modern viewpoint, indeterminate equations that involve several variables or unknowns raised to positive integer values where we search for integer solutions (or more generally, rational number solutions) are now called *Diophantine equations*, and the study of such equations is called Diophantine analysis. It is an extremely rich field.

For our purposes here, because we round off averages in each assignment category, the scores for homework, tests, and the final exam are restricted to nonnegative integer values from 0 to 100. This means that looking at the equation $0.20x + 0.60y + 0.20z = 80$ this way qualifies it as a Diophantine-like equation. Mathematicians call such equations where the variables are all raised to the first power *linear* Diophantine equations.[4] To safely remain in the Diophantine framework, we must consider only those values (of x, y, and z) that combine to be exactly 80, and not those that combine to yield values that round off to 80 (see previous endnote).

Let's think about this. As there are many solutions that satisfy this equation, together they form a numerical symphony—a collective of solutions if you will. Now the question becomes, is there an algebraic way to capture this ensemble?

Without going into details, algebraic maneuvers for this type of equation show that we can find combinations that work from the following relationship:

final exam score = 400 − (homework average) − 3(test average)

or

$$z = 400 - x - 3y.$$

Here, x, y, and z are again nonnegative integers that must be greater than or equal to 0 and less than or equal to 100. Recall that the lower and upper bounds on the values are due to the fact that they represent percentages out of a total with the best score corresponding to 100 (getting everything correct) and the worst corresponding to 0 (getting everything wrong by bad work or through omission).

To see how the relationship $z = 400 - x - 3y$ works, we are free to pick allowable average scores for homework (x) and tests (y) and then plug them into the formula to see what the final exam score (z) would need to be to make the overall course average work out to be exactly 80.

Let's say that a person got a 100 on all of the homework and a 68 average on the tests; what would the final exam score need to be to make it work? Plugging $x = 100$ and $y = 68$ into the formula yields

$$z = 400 - 100 - 3(68) = 400 - 100 - 204 = 300 - 204 = 96.$$

The combination of $x = 100$, $y = 68$, and $z = 96$ should result in a course average of 80. Verification yields

$$0.20(100) + 0.60(68) + 0.20(96) = 20 + 40.8 + 19.2 = 20 + 60 = 80.$$

If a student had an average of 85 on their homework and an 84 on their tests, then we would have $x = 85$ and $y = 84$. To find what final exam score is needed, we plug these values into $z = 400 - x - 3y$ and obtain

$$z = 400 - 85 - 3(84) = 400 - 85 - 252 = 315 - 252 = 63.$$

The combination of $x = 85$, $y = 84$, and $z = 63$ should result in an average of 80. Verification yields

$$0.20(85) + 0.60(84) + 0.20(63) = 17 + 50.4 + 12.6 = 17 + 63 = 80.$$

You can verify that the five solutions in the table are all related through the formula, as well. Many more combinations can be generated, as we did with the last two examples and as you may do by further experimentation.

TWO MORE INDETERMINATE SITUATIONS

COMBINATIONS OF DAYS

Let's next consider the equation $28x + 30y + 31z = 365$. Are there any positive integers that satisfy this equation? Because we are looking for integer solutions, this becomes a Diophantine situation. We will confine our attention to only positive integers, meaning that 0 is not an allowable value for x, y, or z.

Before seeing if there are any solutions, let's first see if we can give this equation an interpretation. If we look closely at the equation, we can observe that the values or coefficients attached to each variable correspond to the number of days for some months out of the year. So, if we let x = number of months with 28 days, y = number of months with 30 days, and z = number of months with 31 days, then the question becomes, are there any combinations of 28-day, 30-day, and 31-day months such that they add up to exactly 365 days for the year?

We know that there is at least one set of values that work, namely the situation that we use in our current calendar (for non-leap years), as shown in the following table:

Number of Days	Month Names	Number of Months	Unknown Value
28	February	1	$x = 1$
30	April, June, September, November	4	$y = 4$
31	January, March, May, July, August, October, December	7	$z = 7$

Placing these x, y, and z values into the expression on the left-hand side yields

$$28(1) + 30(4) + 31(7) = 28 + 120 + 217 = 148 + 217 = 365.$$

Thus, $x = 1$, $y = 4$, and $z = 7$ qualifies as one solution to this Diophantine equation.

Are there any other 28-day, 30-day, and 31-day month configurations that combine to equal 365 days in a year? It turns out that, unlike the situation with course grades, there is only one other combination that works—that includes each length of month at least once—namely, the one with $x = 2$, $y = 1$, and $z = 9$. This corresponds to two months that are 28 days long, one month that is 30 days long, and nine months that are 31 days in length. Plugging these values into the left-hand side of the equation gives $28(2) + 30(1) + 31(9)$, which becomes after multiplication $56 + 30 + 279 = 86 + 279 = 365$.

Interestingly, this second situation still corresponds to a 12-month year, which we didn't insist upon. There are only two 12-month combinations that work. It turns out, in fact, that even if we allow for more or less than 12 months in a year, no other combination of 28-, 30-, and 31-day months will work for a 365-day year.

As with much of this chapter, the details demonstrating this fact are beyond the scope of this book.[5]

ADULT AND CHILDREN'S MUSEUM TICKETS

Consider a Natural History Museum in which admission tickets for adults cost $12 each and for children $8 each. If a total of $400 is available for purchasing tickets, are there any combinations of adults and children that can attend where all of the money is exactly spent with nothing left over? We assume that at least one adult and one child must be in attendance for any of the combinations.

Proceeding algebraically, we need to represent the unknown number of adults and number of children each by a specific letter. We let x = number of adults and y = number of children. We can treat the setup similarly to what we did with coins as packaged currency in

Chapter 5, where here each adult can be considered to have a value of $12, whereas each child has a value of $8. The dollar value of the varying combinations of adults and children who attend the museum can be represented by

$$12(\text{number of adults}) + 8(\text{number of children}),$$

which translates to

$$12x + 8y.$$

If 5 adults and 14 children attend, the cost would be $12(5) + 8(14) = 60 + 112$ or $172. This amount, of course, doesn't exactly exhaust the $400.

To algebraically set up the situation of interest, we must represent the condition "total adult cost + total child cost = 400" by an equation. This readily translates to $12x + 8y = 400$.

Because the number of adults and children must be equal to positive integers (we can't have 3.7 adults or 7.4 children), we are looking at this equation through the Diophantine lens.

We list a couple of solution scenarios in the following table:

Number Going to the Museum	Variable Settings	Cost = $12x + 8y$	Total Cost
Adults = 18 Children = 23	$x = 18$ $y = 23$	$12(18) + 8(23) = 216 + 184 = 400$	$400
Adults = 8 Children = 38	$x = 8$ $y = 38$	$12(8) + 8(38) = 96 + 304 = 400$	$400

An algebraic formula exists that can be used to generate all of the possible combinations of adults and children that work. In words and symbols, we have

$$\text{number of children} = 50 - 1.5(\text{number of adults})$$

or

$$y = 50 - 1.5x.$$

A demonstration of how this formula can be obtained by solving for y in the equation $12x + 8y = 400$ is shown in Appendix 5.

There are restrictions on the x, namely that it can't be a number that causes y to have a negative or fractional value. This means that x (the number of adults) can't be larger than 32. If there were 33 adults, then y would be a fraction, and 34 adults gives a value of -1 for y. It also must be an even number as only an even number multiplied by 1.5 will yield an integer.

On the other side of the equation, the number of children (y) can be either even or odd as the two examples in the table show, and it ranges from a low of 2 to a high of 47. It turns out that there are a total of 16 positive integer solutions to this equation or a total of 16 different adult-child combinations that work.

If we form the doublet (number of adults, number of children) = (x, y), then the solution combinations that work are these 16 doublets:

(2, 47); (4, 44); (6, 41); (8, 38); (10, 35); (12, 32); (14, 29); (16, 26); (18, 23); (20, 20); (22, 17); (24, 14); (26, 11); (28, 8); (30, 5); (32, 2).

Many more scenarios open up, of course, if we remove the criteria that the budgeted $400 must be exactly spent and simply say that we can spend no more than $400.

A SECOND-DEGREE EQUATION

Let's now examine an issue involving those special positive integers that are perfect squares. A perfect square integer is one that is equal to another integer multiplied by itself (another integer squared).[6] Thus, 9 is a perfect square because it is equal to the integer 3 times itself (3×3 or 3^2), and so is 16, which equals 4 times itself (4×4 or 4^2). The number 12, however, is not a perfect square as no integer times itself equals 12.

The integers between 1 and 200 that are perfect squares are 1, 4, 9, 16, 25, 36, 49, 64, 81, 100, 121, 144, 169, and 196. Adding some of these perfect squares together in pairs reveals the following:

Label	Perfect Pair	Sum
A	1, 4	$1 + 4 = 5$
B	4, 9	$4 + 9 = 13$
C	9, 16	$9 + 16 = 25$
D	25, 36	$25 + 36 = 61$
E	36, 64	$36 + 64 = 100$
F	49, 81	$49 + 81 = 130$
G	25, 144	$25 + 144 = 169$
H	16, 169	$16 + 169 = 185$

Five of the sums (A, B, D, F, H) add up to values that are not in the list of perfect squares, but three of them (C, E, G) do add up to values that are themselves perfect squares. So, in these three cases we have the situation where one perfect square plus a second perfect square adds up to yield a third perfect square.

Are there other triplets of three positive integers [such as (9, 16, 25), (36, 64, 100), and (25, 144, 169)] that coordinate this way, that is, three perfect squares such that the largest is equal to the other two added together? This question smells a bit like the other problems involving indeterminacy where we sought to find sets of three integer average scores, sets of three types of months, or combinations of adults and children that blended together, respectively, to yield a course average of 80, a total of 365 days, or a sum of $400 spent.

The difference this time is that instead of trying to find a group of three numbers that give the same *value*—same course average, same number of days, or same amount of money—we are trying to find a group of three integers—each with the special property that they are perfect squares—such that they satisfy the same *relationship* with each other: that two of them add up to give the third. Put another way, in the present case there is not a stable value like the course average of 80 (or 365 days or $400), but rather the stability is encoded in the property of the numbers being perfect squares and the relationship they satisfy.

That is, the numbers (9, 16, 25) are all different from (36, 64, 100) and (25, 144, 169), yet in each of the three cases, the property that each number is a perfect square such that the smaller two add up to the third, is satisfied.

There are many equivalent ways to phrase this inquiry. We list a few here:

1. Find positive integers that are perfect squares that can be split into a sum of two other positive integers that are also themselves perfect squares.
2. Find sets of three perfect squares (positive integers) such that one of them is the sum of the other two.
3. Find situations that allow us "to divide a square into two other squares" (Fermat's language—see this chapter's epigraphs).

We can try to find answers to this inquiry by trial and error, but that is a daunting process because the majority of sums of two perfect squares don't add up to a third perfect square. The 14 perfect squares that are less than 200 yield 105 unique pairs of two numbers added together, where unique means that we only count once the sums that are identical when the numbers are reversed. For example, this means that we would only count $1 + 4$ and $4 + 1$ as one unique pair instead of two pairs because the numbers involved in both sums are the same and addition is commutative. Similarly, the four pairs $25 + 36$, $36 + 25$, $49 + 64$, and $64 + 49$ would only count as two unique pairs.

Out of the 105 unique pairs, only four ($9 + 16$, $36 + 64$, $25 + 144$, and $81 + 144$) add up to a perfect square: 25, 100, 169, and 225, respectively. This amounts to less than 4% of the combinations of the first 14 perfect squares meeting the criteria.

There are many more perfect squares than these first 14, so the difficulty of finding pairs that give us what we want only magnifies.

Can algebra help us out here? And if so, how do we even begin to deploy it in this case?

The situation here involves three special numbers, two of which add together to yield the third, larger number. In the algebraic mindset, we can treat each of these numbers as a variable or unknown. Doing so means that we would then represent each by a letter. Let's choose the usual suspects: x, y, and z. Continuing, we will let $x =$ the first perfect square integer, $y =$ the second perfect square integer, and $z =$ the third perfect square integer, such that $x + y = z$.

This representation is problematic, however: Before we can engage it, we must make sure that x, y, and z each have the special property of being perfect square integers. How can we be sure? This is really tantamount to the trial-and-error case because we must already know this information about the three numbers before we even engage the algebraic equation; but if we are required in every case to know the information up front, then the algebra is not really helping us.

We need a different plan. As a start, let's rewrite the perfect squares less than 200 in exponential form as follows:

- **Perfect square form:** 1, 4, 9, 16, 25, 36, 49, 64, 81, 100, 121, 144, 169, 196.
- **Exponential form**[7]: 1^2, 2^2, 3^2, 4^2, 5^2, 6^2, 7^2, 8^2, 9^2, 10^2, 11^2, 12^2, 13^2, 14^2.

The following table shows how the four situations where two perfect squares add to a third perfect square translate into exponential form:

Perfect Square Form	Exponential Form
9 + 16 = 25	$3^2 + 4^2 = 5^2$
36 + 64 = 100	$6^2 + 8^2 = 10^2$
25 + 144 = 169	$5^2 + 12^2 = 13^2$
81 + 144 = 225	$9^2 + 12^2 = 15^2$

Now, if we look at the equation $x^2 + y^2 = z^2$, then the bases of the exponents in the second column of the previous table can be viewed as being integer solutions to this equation, as demonstrated in the following:

Exponential Form	x	y	z	Integer Solution to $x^2 + y^2 = z^2$?
$3^2 + 4^2 = 5^2$	3	4	5	YES
$6^2 + 8^2 = 10^2$	6	8	10	YES
$5^2 + 12^2 = 13^2$	5	12	13	YES
$9^2 + 12^2 = 15^2$	9	12	15	YES
$3^2 + 5^2 = 6^2$	3	5	6	NO

Note that the bottom row of values $x = 3$, $y = 5$, and $z = 6$ (which upon substitution into $x^2 + y^2 = z^2$ give $3^2 + 5^2 = 6^2$) do not represent a solution of the equation, because the left-hand side gives $9 + 25 = 34$ but the right-hand side gives 36; and these numbers are not equal. Try a few random values for x, y, and z yourself, and you will see that most of them won't be solutions.

If we use the exponential form for our new game plan, we can restate all three of the perfect square questions as: "Find three positive integers (x, y, and z) such that they satisfy the equation $x^2 + y^2 = z^2$."

In this form, it becomes more transparent what we seek, and we are free to choose various integer inputs of x, y, and z and place them into the equation to see if they work. That is, this equation now involves the squares of numbers, meaning that no matter what integer values of x, y, and z we initially pick, once we place them into the equation, we get the values x^2, y^2, and z^2, and then we know that we are dealing with three perfect squares.

Consequently, perfect squares are automatically generated from the inputs as we engage the equation. The only thing left to decide is whether or not they satisfy the equation. This is different from using the equation $x + y = z$ whose form does not make transparent what we seek, moreover requiring us to preselect perfect squares before engaging the equation. The automatic generation of perfect squares in the equation $x^2 + y^2 = z^2$ makes it the more natural and operational setting for the question of sums of perfect squares.

Now that we are seeking positive integer solutions, we are in the Diophantine outlook and this equation becomes a Diophantine equation, this time of the second degree. The second degree comes from the fact that the variables are squared or raised to the second power. In this degree language, the linear Diophantine equations discussed in the previous two sections can be thought of as being Diophantine equations of first degree, because the variables in each are raised only to the first power.

Before looking at possible solutions to the equation involving perfect squares, let's first see if we can give it a more familiar interpretation, similarly to what we did in the last section with the equation $28x + 30y + 31z = 365$.

It turns out that we can by relating it to one of the most famous mathematical results of all time: the Pythagorean Theorem. The Pythagorean Theorem expresses a relationship between the sides of a right triangle, where a right triangle is one that contains a 90-degree angle (also called a right angle).

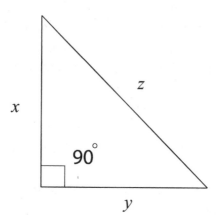

The Pythagorean Theorem: For a right triangle with legs x and y and hypotenuse z, the following relationship holds: $x^2 + y^2 = z^2$.

The question about the sum of perfect squares can translate to: "Find right triangles such that the lengths of all three of the sides are positive integers." The lengths of any three sides that satisfy this condition are called Pythagorean triples or triplets. We have mentioned four so far. Placing these in geometric form yields the following right triangles:

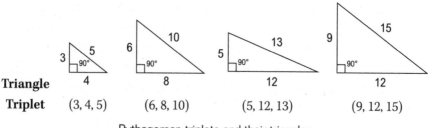

Pythagorean triplets and their triangles

Fortunately, it turns out that there are algebraic ways to generate Pythagorean triples quite easily and to our heart's content.

We will use new varying quantities—designated by m and n—to state the relationships (once again, the demonstration on how these are obtained is beyond the scope of our discussion):

$$x = m^2 - n^2, \quad y = 2mn, \quad z = m^2 + n^2.$$

Let's see how this works. If we choose $m = 2$ and $n = 1$, this will yield the following:

$$
\begin{array}{ccc}
x = m^2 - n^2 & y = 2mn & z = m^2 + n^2 \\
\downarrow & \downarrow & \downarrow \\
x = 2^2 - 1^2 & y = 2(2 \times 1) & z = 2^2 + 1^2 \\
\downarrow & \downarrow & \downarrow \\
x = 4 - 1 & y = 2(2) & z = 4 + 1 \\
\downarrow & \downarrow & \downarrow \\
x = 3 & y = 4 & z = 5
\end{array}
$$

This produces the Pythagorean triple (3, 4, 5), which we already know. Let's now choose $m = 7$, $n = 5$, and find x, y, and z:

$$
\begin{array}{ccc}
x = m^2 - n^2 & y = 2mn & z = m^2 + n^2 \\
\downarrow & \downarrow & \downarrow \\
x = 7^2 - 5^2 & y = 2(7 \times 5) & z = 7^2 + 5^2 \\
\downarrow & \downarrow & \downarrow \\
x = 49 - 25 & y = 2(35) & z = 49 + 25 \\
\downarrow & \downarrow & \downarrow \\
x = 24 & y = 70 & z = 74
\end{array}
$$

This yields the Pythagorean triple (24, 70, 74), which has not been mentioned. Note that $24^2 + 70^2 = 74^2$, which as perfect squares is 576 + 4900 = 5476.

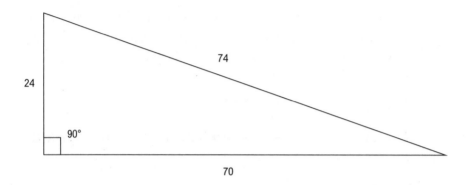

The following table gives a list of several more Pythagorean triples that can be generated from values for m and n,[8] where it is always the case that $x^2 + y^2 = z^2$:

m, n	Pythagorean Triple x, y, z	Perfect Square Form x^2, y^2, z^2
2, 1	3, 4, 5	9, 16, 25
3, 2	5, 12, 13	25, 144, 169
3, 1	6, 8, 10	36, 64, 100
4, 3	7, 24, 25	49, 576, 625
4, 1	15, 8, 17	225, 64, 289
5, 4	9, 40, 41	81, 1600, 1681
6, 5	11, 60, 61	121, 3600, 3721
6, 1	35, 12, 37	1225, 144, 1369
7, 6	13, 84, 85	169, 7056, 7225
8, 7	15, 112, 113	225, 12544, 12769
8, 1	63, 16, 65	3969, 256, 4225
9, 8	17, 144, 145	289, 20736, 21025
10, 9	19, 180, 181	361, 32400, 32761
10, 1	99, 20, 101	9801, 400, 10201
5, 2	21, 20, 29	441, 400, 841
14, 1	195, 28, 197	38025, 784, 38809
15, 14	29, 420, 421	841, 176400, 177241
20, 19	39, 760, 761	1521, 577600, 579121
20, 1	399, 40, 401	159201, 1600, 160801

The variables m and n cannot generate all of the Pythagorean triples out there. For instance, no combination of integer values for m and n can generate (9, 12, 15). However, this particular triplet is an integer multiple of a more basic triplet that can be generated by m and n, namely (3, 4, 5). By integer multiple, we mean that (9, 12, 15) can be obtained from (3, 4, 5) by multiplying each term by the same integer, in this case 3: that is, $3 \times (3, 4, 5) = (3 \times 3, 3 \times 4, 3 \times 5) = (9, 12, 15)$. It turns out to be the case, in general, that if m and n cannot generate a certain triple, then that triplet will turn out to be an integer multiple of a more basic one that can be generated by m and n.

Thus, algebra once again is able to decisively help us in answering the question of finding sets of three perfect squares such that one of them is the sum of the other two.

Through the m and n formulas, we are able to generate with ease and precision as many sets of three numbers that work as we like. Remember, there are a sea of possibilities to choose from, and the overwhelming majority of those choices, if made by guesswork, will not work.

Geometry serves as a critical aid here, too, by providing us with conceptual insight through recognizing the problem as one that can be related to the sides of right triangles, which in turn allows us to involve the Pythagorean Theorem.

In the sense of "The Star-Spangled Banner" metaphor, we can think of the right triangle interpretation and the perfect squares interpretation as different renditions of the algebraic song given by the equation $x^2 + y^2 = z^2$: renditions that may assist us—through the shuttling back and forth between viewpoints—in better understanding the problem and the equation.

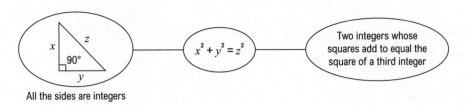

All the sides are integers

A BABYLONIAN SURPRISE

The different renditions of the equation $x^2 + y^2 = z^2$ are also able to assist us in providing interesting commentary on an important historical question: Were others aware of the Pythagorean Theorem before the time of Pythagoras (ca. 570–490 BC)?

Surprisingly, through the analysis of several clay/cuneiform tablets from the days of ancient Babylon (ca. 1800 BC) over 1000 years before the time of Pythagoras, it appears that the Babylonians were already familiar with the result. One of the cuneiform documents that gave twentieth-century mathematicians some of that initial confidence is called Plimpton 322. It has a table that, among other things, lists two of the three sides of 15 right triangles that have integer-valued lengths. The sides included on the tablet are the shortest side of the right triangle and the longest side. In terms of our list in the table of m and n values, these would correspond to the short leg (x) and the diagonal side or hypotenuse (z). For example, the Pythagorean triplets (3, 4, 5) and (7, 24, 25) would appear on Plimpton 322 as showing only (3, 5) and (7, 25), respectively. The side whose length is in the middle would not have its value included.

There is a great deal of mystery and disagreement among mathematical historians surrounding the tablet because none of the smaller Pythagorean triples, such as those from our table giving m and n values, are included. The list of values included on Plimpton 322 are given in the following table using modern-day numerals:

List of Ratios	Short Leg (x)	Hypotenuse (z)	Entry #	m	n
	119	169	1	12	5
	3367	4825	2	64	27
	4601	6649	3	75	32
	12,709	18,541	4	125	54
	65	97	5	9	4
	319	481	6	20	9
	2291	3541	7	54	25
Plimpton values not reproduced here	799	1249	8	32	15
	481	769	9	25	12
	4961	8161	10	81	40
	45	75	11	–	–
	1679	2929	12	48	25
	161	289	13	15	8
	1771	3229	14	50	27
	56	106	15	9	5

Values on Plimpton 322 tablet (in our modern-day numerals)

Note that the shaded cells listing the accompanying m and n values are not on the Plimpton 322 tablet. In the last row, the sides generated from $m = 9$ and $n = 5$ are not what mathematicians call a primitive triple. Also note that the tablet was written in cuneiform using base 60 or sexagesimal notation.[9]

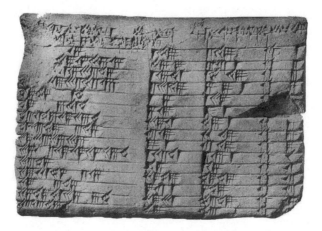

Actual Plimpton 322 clay tablet [Photo from *Mathematical Cuneiform Texts* (Neugebauer and Sachs 1945), permission provided courtesy of the American Oriental Society]

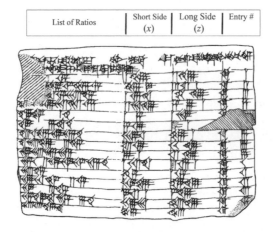

List of Ratios	Short Side (x)	Long Side (z)	Entry #

Drawing of Plimpton 322 [Image from (Robinson 2002), provided courtesy of Eleanor Robson; the column headings in box were added by the author]

$$1 = \Gamma \; ; \; 2 = \Gamma\Gamma \; ; \; \ldots \; 10 = \langle \; ; \; 11 = \langle\Gamma \; \ldots$$

Decimal cuneiform relationships

The smallest of the fifteen triplets included on Plimpton 322 (entry 11) corresponds to a right triangle with sides $x = 45$, $y = 60$, and $z = 75$. On the table, only 45 and 75 are included. This triplet cannot be generated by m and n values, but is 15 times the triplet (3, 4, 5), which can be generated by the parameters $m = 2$ and $n = 1$: $x = 15 \times 3$, $y = 15 \times 4$, and $z = 15 \times 5$.

The largest value on the Plimpton 322 tablet (entry 4) corresponds to the triplet (12709, 13500, 18541), where only the values 12,709 and 18,541 are included. These values give the following relatively large right triangle:

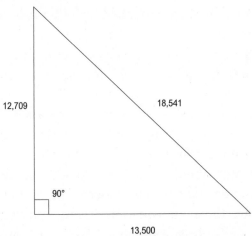

These numbers satisfy the relationship $x^2 + y^2 = z^2$: $12709^2 + 13500^2 = 18541^2$, which in perfect squares form is $161518681 + 182250000 = 343768681$. Note that these x, y, and z values can be generated from the m and n formulas by setting $m = 125$ and $n = 54$.

Why the Babylonians chose such large and sporadic values to list in the Plimpton 322 tablet has mystified mathematicians since the 1940s, when the fact that the rows in the tablet list two of the values in 15 different Pythagorean triplets was discussed by the Austrian mathematical historian Otto Neugebauer and American Assyriologist Abraham Sachs.[10] It is still an unsettled question as to what purposes the Babylonians had in mind in composing Plimpton 322.

In 2001, the English mathematical historian Eleanor Robson offered an explanation, suggesting that the values in the tablet were used for educational purposes.[11] For a 2002 article on this theme, she was awarded the Paul R. Halmos–Lester R. Ford Award for expository excellence from the Mathematical Association of America in 2003.[12]

In 2017 and 2021, Plimpton 322 made a splash in the news once again as mathematicians in Australia resurrected earlier claims that the values in the tablet were used for making trigonometrical calculations, but other scholars and historians, including Robson, question this interpretation.[13]

Robson's consistent stance has been that we in our era must be careful not to read too much into the historical record based on our modern-day viewpoints, and that the intentions of the ancient authors were often very different from ours. Many other historians share this perspective as well.

To get a sense of the historians' reasons for urging caution, let's go back to "The Star-Spangled Banner" (or algebraic songs) metaphor from Chapter 7, where, if you recall, the equation $16x + 10 = 106$ has multiple interpretations. Now, let's imagine that each of the five interpretations were in vogue during a different period of history, with the one concerning finding an unknown number (problem 1) being the outlook of ancient mathematicians in 1800 BCE and the one involving power tools (problem 4) being our modern outlook. Because they both ultimately involve the same equation, if after translating an ancient text, we found the equivalent of the modern equation $16x + 10 = 106$, we might conclude that the ancient mathematicians knew something about power tools. But this conclusion, of course, would be very wrong.

The simplicity of this analogy makes it a bit of a stretch, but hopefully it illustrates the point that sometimes the very strength of mathematics (the universality of its rules and relationships along with the ensemble of interpretations it allows) can from time to time lead us astray in making accurate assessments of the intentions of the ancients. This sometimes may lead us to underestimate their aims, but it can lead to sometimes overestimating them, as well.

All of this means that more than mathematical formulas are required to really understand what was going on in an earlier era. We must also look at the context of the times, what the authors themselves say, and other documents when available. Given the great length of time from then to now—and the dearth of documents in ancient languages—this can be a monumental, oftentimes impossible, task.

Whatever the reality in terms of the Babylonian authors' intentions, there is no doubt that the Plimpton 322 tablet is cuneiform and that values on it can be interpreted as being two of the three sides of a right triangle with integer sides. This combined with other cuneiform documents makes this segment of Mesopotamian mathematics interesting in its own right and demonstrates that mathematicians were utilizing the rule associated with Pythagoras of Samos hundreds of years before he flourished. This is something that we have only been made keenly aware of since the middle of the twentieth century.[14]

FERMAT'S LAST THEOREM

The problems with perfect squares can be naturally extended to other powers. Let's next consider perfect cubes.

A perfect cube integer is one that is equal to another integer multiplied together three times (another integer cubed). Thus, 8 is a perfect cube because it is equal to the integer 2 multiplied together three times ($2 \times 2 \times 2$ or 2^3), and so is 125, which equals 5 multiplied together three times ($5 \times 5 \times 5$ or 5^3). The integers that are perfect cubes from 1 to 200 are 1, 8, 27, 64, and 125.

Now let's take some perfect cubes and add them together in pairs:

Label	Pair of Perfect Cubes	Sum
A	1, 8	$1 + 8 = 9$
B	8, 27	$8 + 27 = 35$
C	8, 64	$8 + 64 = 72$
D	27, 64	$27 + 64 = 91$
E	64, 64	$64 + 64 = 128$
F	1, 125	$1 + 125 = 126$
G	27, 125	$27 + 125 = 152$
H	64, 125	$64 + 125 = 189$

Notice that none of the sums add up to values that are perfect cubes, nor do any of the other pairs of sums from these five numbers (15 unique pairs in total). In fact, it turns out that even if we extend the perfect cubes to include other values such as 216, 343, 512, 729, and 1000, we will never find a sum of two that add up to another perfect cube.

A total shutout from the infinitely many combinations possible.

Put in exponential language, we have that the equation $x^3 + y^3 = z^3$ has no positive integer solutions at all! And not only is this shutout true for the cubes (or third-degree equation), but it is also true for the fourth-degree equation $x^4 + y^4 = z^4$. Our discussions so far summarize as follows:

- There are **infinitely many** combinations of three positive integers $(x, y, \text{ and } z)$ that satisfy the equation $x^1 + y^1 = z^1$, or equivalently $x + y = z$. Examples include $(x = 7, y = 10, z = 17)$ and $(x = 432, y = 68, z = 500)$.
- There are **infinitely many** combinations of three positive integers $(x, y, \text{ and } z)$ that satisfy the equation $x^2 + y^2 = z^2$. Examples include $(x = 3, y = 4, z = 5)$ and $(x = 7, y = 24, z = 25)$.
- There are **no** combinations of three positive integers $(x, y, \text{ and } z)$ that satisfy the equation $x^3 + y^3 = z^3$.
- There are **no** combinations of three positive integers $(x, y, \text{ and } z)$ that satisfy the equation $x^4 + y^4 = z^4$.

It turns out that this total shutout of integer solutions holds not only for the scenarios of third and fourth powers but for all higher integer powers as well. We will use the letter n as a parameter to restate all of these scenarios in one single big algebra equation as follows:

There are **no** combinations of three positive integers $(x, y, \text{ and } z)$ that satisfy the equation $x^n + y^n = z^n$ for $n \geq 3$ (n is an integer greater than or equal to 3).

or

The equation $x^n + y^n = z^n$ has no positive integer solutions for $n \geq 3$ (n is an integer greater than or equal to 3).

Though it is common to read it in the singular, the equation $x^n + y^n = z^n$ really represents infinitely many equations (one for each value of n).

This represents another of the very famous storylines in mathematics: *Fermat's Last Theorem*. Pierre de Fermat lived in the 1600s and proved many interesting results about relationships amongst the positive integers. He was a regular correspondent with other mathematicians throughout his life, often sharing the results he proved as challenge problems to them.

Though his efforts in number theory are what he is most famous for, his contributions in analytic geometry, the infinitesimal calculus, physics, and probability are immense and often underrated. In fact, Sir Isaac Newton acknowledged that he received some of his ideas and inspiration for calculus directly from Fermat's work.[15] Fermat was truly one of the most remarkable mathematicians of the seventeenth century.

At some point during his research, Fermat jotted down in the margins of his copy of Diophantus' book, *Arithmetica*, that he was able to demonstrate why the equations $x^n + y^n = z^n$ all had no solutions for integers n larger than 2, but that he didn't have enough space in the margins to show it (see this chapter's epigraphs). And ultimately, he never did show it in his writings.

There are many intriguing items surrounding this story, some mathematical and some historical. Two of the most fascinating are as follows:

1. Why do the equations go from having infinitely many solutions when $n = 1$ and $n = 2$ to having no solutions at all for n larger than 2 ($n = 3$, $n = 4$, $n = 5$, and so on)?

    That is, what is so special about the two lower powers (especially the squares) and so intractable about higher powers such that the distribution of squares allows for infinitely many solutions, whereas the distribution of cubes, fourth powers, fifth powers, and higher don't allow any? Why not at least a few solutions for these? One might expect a gradual reduction in the number of solutions as the powers get higher—and the spacing between numbers with that particular power property increases—but why the sharp cliff from infinitely many solutions to none at all, in going from $n = 2$ to $n = 3$?

2. Did Fermat actually find a way to demonstrate the general result, and if so why didn't he share it as a challenge to others or write it down somewhere in his papers?

Mathematicians have found the answer to the first question to be quite intricate and related to a chain of deep results. The last link in that chain was supplied in late 1994 (and published in 1995) by the celebrated mathematician Andrew Wiles (assisted by his former student Richard Taylor), and the proof of Fermat's Last Theorem is said to come out as a consequence.

Regarding the second question, most mathematicians doubt that Fermat actually found a general answer. They base this in part on two things:

1. No airtight proof was ever found by the many excellent mathematicians who have worked on the theorem from the time of Fermat until the 1990s. This spans a period of over 300 years.
2. The arguments developed by Wiles to prove the theorem were quite long (over 100 journal pages in length) and used complex mathematics that was only developed in the twentieth century and unknown to Fermat.

The fact that Fermat didn't send it out as a challenge to others also suggests that he may have realized that there were errors in his argument and thus did not in the end have the result that he privately claimed in the margins of his copy of *Arithmetica*.

CONCLUSION

Over the years, many have questioned the worth of pursuing scientific knowledge out of sheer curiosity and imagination alone, only later to have new emergent outlooks and technologies slam the door on that skepticism.

Electricity and magnetism present a case in point. The sophisticated electrified world that we inhabit today had its origins in the curiosity of the early investigators of electricity and magnetism in the eighteenth and nineteenth centuries, men such as Charles-Augustin de Coulomb, Alessandro Volta, André-Marie Ampère, Hans Christian Ørsted, Michael Faraday, William Thomson (Lord Kelvin), and James Clerk Maxwell. At the time, it was not obvious what these investigations would grow up to be or if they would even grow up to be anything applicable at all. But grow up they did, and now it is hard to imagine a world without the technologies that spewed forth out of this knowledge.

The investigations of physicists into the structure of the atom offer another case, where our modern world of televisions, cell phones, and computers would not be possible without the curiosity-driven discoveries of these late nineteenth- and early twentieth-century scientists.

Curiosity is one of scientists' most valuable tools. And even when discoveries find no known application, they provide an important background canvas for situating those discoveries that do have a direct impact. It is similarly so with mathematicians. Much of the theoretical advance in science and technology has and continues to be underwritten by mathematics that was originally created more out of simple curiosity about the subject itself rather than with an eye to direct utility.

In considering integer solutions to indeterminate equations, mathematicians embarked on a curiosity-laden journey that demonstrates some of what can happen when free inquiry interacts with imagination, mathematical analysis, and circumstance. A bit of this is in evidence from our basic considerations here in this chapter. Let's discuss a few of these.

COMPLEMENTARY PERSPECTIVES

The Diophantine equation $x^2 + y^2 = z^2$ connects to the Pythagorean Theorem for right triangles. Let's look at a couple of things that can occur in the mathematical interplay of ideas from such a connection.

If we look at this equation as only involving the sum of two perfect square integers adding up to equal a third perfect square integer, we will be happy when we find values for x, y, and z that work and throw away the ones that don't.

However, because the equation also connects to right triangles, we can look at the situation geometrically. When $x = 1$ and $y = 1$, yielding $1^2 + 1^2 = z^2$, and we discover that we can't find any integer values for z such that $2 = z^2$, we can't simply throw away these values for x and y. There exists a right triangle with side lengths $x = 1$ and $y = 1$:

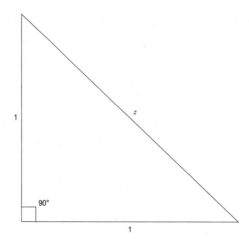

The circumstances now involve an actual hypotenuse, z, that has an actual length. And the question becomes, if this length is not an integer, then what is it? An answer is now demanded. The length, of course, is $z = \sqrt{2} \approx 1.41421\ldots$ This length is not an integer, nor is it even the ratio of two integers (common fraction). It presents us with an example of an entirely new category of numbers called *irrational numbers*. These numbers have infinitely long decimal representations that don't repeat in a pattern of digits.

So, looking at this problem from the standpoint of integers alone would lead us to say that it has no solution and then be done with it. But looking at it from the other interpretation—as sides of a right triangle— we can't ignore it. The need to answer the question of what the length of z is eventually leads to the discovery of a whole new class of numbers.

If we look at the equation $x^2 + y^2 = z^2$ only from the geometrical view-point, we may see it as a two-dimensional space problem. This could lead us to naturally extend it to three dimensions, which yields $x^3 + y^3 = z^3$, and then seek integer solutions there. But because we are in the geometric viewpoint, and given that the geometrical objects we can easily imagine or visualize consist of either zero dimension (points), one dimension (lines, curves, etc.), two dimensions (rectangles, trian-gles, etc.), or three dimensions (cubes, spheres, etc.), we may stop at this third-degree equation, thinking that there are no more physical dimensions to consider.

However, if we look at it from the exponential powers viewpoint, there is nothing to keep us from considering what the fourth power or fifth power versions of this equation look like and so on, leading us ultimately to the equation $x^n + y^n = z^n$.

This type of thing happens frequently in mathematics where one representation or viewpoint of a phenomena—or algebraic expression—opens up a whole world of new possibilities to which a second representation or viewpoint is totally blind. In the cases discussed here, both viewpoints—geometrical and integer—yield insights that the other respective viewpoint doesn't naturally offer.

CREATION OF NEW MATHEMATICS

Larger problems are often identified from generalizations of questions that arose out of smaller, particular situations. The search then for commonalities in these larger problems can often lead to dramatically new insights—entire new areas of mathematics even. Two examples of this flow from our discussions in this chapter.

FERMAT'S EXTENSION OF THE DIOPHANTINE EQUATION $X^2 + Y^2 = Z^2$:

When Fermat extended his investigations to higher powers, initiating a search for positive integer solutions to $x^n + y^n = z^n$, he was led to his conjecture for values of $n \geq 3$, which become known as his famous Last Theorem.

The attempts to prove this theorem, or show gaps in proposed proofs offered, ultimately spawned a large body of new mathematics whose importance far surpasses the importance of proving the original question itself.

The Scottish-American mathematician Eric Temple Bell had exactly this type of thing in mind when he shared (in a statement he attributed to mathematician Felix Klein): "Choose one definite objective and drive ahead toward it. You may never reach your goal, but you will find something of interest along the way."[16]

HILBERT'S EXTENSION OF SOLUTIONS TO DIOPHANTINE EQUATIONS IN GENERAL (TENTH PROBLEM): In looking at the basic Diophantine equations discussed here, there was a great difference in the form of

the solutions to linear Diophantine equations (such as $12x + 8y = 400$) and the solutions to the second-degree equation given by the Pythagorean Theorem ($x^2 + y^2 = z^2$). There is also wide variation in how the solutions for each were obtained. This is also true of other Diophantine equations such as the following two:

- $y^2 - 17 = x^3$.

  Here are three integer solutions (out of many):

$$(x = 2, y = 5), \quad (x = 4, y = 9), \quad (x = 52, y = 375).$$

- $x^4 - x^2y^2 + y^4 = z^2$.

  Here are three integer solutions (out of many):

$$(x = 5, y = 5, z = 25), \quad (x = 10, y = 10, z = 100), \quad (x = 75, y = 75, z = 5625).$$

The techniques for finding the integer solutions to these two equations are different from the ones we have discussed in detail.

However, there are other Diophantine equations out there that don't have any positive integer solutions at all, such as $x^{10} + y^{10} = z^{10}$, $x^{15} + y^{15} = z^{15}$, and $5x^4 + 7y^6 = 1$.

Thus, Diophantine equations exist that have either no solutions, a small number of solutions, many more (but still finite) solutions, and infinitely many solutions. This means that the question of whether or not a solution exists will vary from Diophantine equation to Diophantine equation. Since the time of Diophantus, mathematicians have come up with various techniques for solving specific Diophantine equations, even determining when many have no solution. As with those we have discussed, the technique used depended on the particular equation involved.

In 1900 during the International Congress of Mathematicians in Paris, the great German mathematician David Hilbert was so bold as to propose that mathematicians find out if it was possible to find one single, all-encompassing, universal method that would work on any and all Diophantine equations, in determining whether they had solutions or not.[17]

During his talk on Wednesday, August 8, Hilbert discussed only ten of his total list of twenty-three interesting questions that he thought twentieth-century mathematicians should pay some attention to. This question concerning Diophantine equations was number ten on his written list and is known as Hilbert's Tenth Problem. It was not discussed in his talk. Hilbert's wording of the problem translated to English summarizes as follows:

> Given a Diophantine equation with any number of unknown quantities and with… [integer]…numerical coefficients: To devise a process according to which it can be determined by a finite number of operations whether the equation is solvable in… [integers].

This question was eventually answered in the negative in 1970—no single universal method exists—by the Russian mathematician Yuri Matiyasevich. Matiyasevich's final conquest of the problem depended critically on the work of three American mathematicians: Martin Davis, Hilary Putnam, and Julia Robinson. Interesting new ideas came out of the techniques used to demonstrate this result, as with Fermat's Last Theorem.

HISTORICAL INTERPLAY

In studying integer solutions to $x^2 + y^2 = z^2$ from the geometric viewpoint, mathematicians were led to the notion of Pythagorean triples, eventually finding ways to generate them at will. Euclid gives a demonstration of how to do this in his famous book, *Elements*, while Viète gives a more symbolic demonstration in his work at the end of the sixteenth century.[18]

As modern linguists began to crack the ancient Cuneiform writing systems of the Sumerians and Babylonians, modern mathematicians—with their crucial assistance—were eventually able to also acquire an appreciation of some of the mathematics that was known to these ancients. This included their work on second-degree situations and right triangles. This would ultimately lead mathematicians to the shocking realization—with evidence—that the Pythagorean property appears to have been known and used hundreds of years before the time of Pythagoras. Without the imagination of both ancient mathematicians and contemporary scholars, this very real possibility might have never been realized.

Information such as this could then in turn enhance the viewpoints of historians on how knowledge was transmitted between the Middle Eastern civilizations (Mesopotamia and Egypt) and Ancient Greece. Such mutualistic relationships often occur between the sciences and other disciplines (e.g., radiocarbon dating).

ALGEBRAIC INTANGIBLES

There is a certain elegance to the manner in which perspectives and generalizations interact to yield new insights and unexpected possibilities. Such unanticipated mathematical convergences and unifications over the broad plain of human ideas certainly qualify, in the view of some, as being aesthetic on a variety of levels.

As such, some artifacts of these transactional flairs certainly are worthy of display, reflection, and discussion. Such exhibition often occurs when a famous mathematical result is proven, or some new mathematical knowledge is discovered and gets written up in the media. Unfortunately, even after reading such articles, it usually turns out that too little is still understood by the lay reading audience, and much of the technical reading audience as well. There is probably no easy way to improve upon this, but it is the merging of ideas in deep and sometimes surprising ways—such as shown in miniature here—that often creates the buzz in the air among the experts in the know.

The benefits that accrue from such curiosity not only can help mathematicians and scientists in their investigations, but may also assist other learners in their own personal journeys by allowing them to add to their own "stores of conceptual fuel."

These are but a few of the kinds of intangibles—from learning algebra—that those partial to the subject sometimes have in mind when they proclaim algebra for all.

In Chapter 11, we illustrate how exploring alternate viewpoints for some of the algebra we have already considered can provide additional insights for both mathematics and science.

11

A Kaleidoscope of Ingredients

> Many things of interest to mathematicians or engineers are very
> complex and have several complicated structures layered on top
> of one another. As we do in other things in life, we like to take things
> apart into more elementary building blocks so we can unravel this
> rich structure.
> To do such unravelling, we view these complex things as a superposition
> of much simpler ones....Decomposing into simple building blocks can be
> done in many different ways in mathematics.
>
> —Ingrid Daubechies, interview for *The Intrepid Mathematician*

Elementary algebra turns out to contain, in the small, examples of
ideas that in the large have played central roles in mathematics and
science. Though many of these roles have their historical origins much
further up the mathematical and scientific food chains (and certainly
are not close to being contained in entirety in these small examples),
the kernel of some of the ideas, upon revisit, are present or reflected
in basic algebra. The aim of this chapter is to discuss a few of those
kernels.

SEPARATING OUT NUMERICAL INTERACTIONS (REPRISE)

In 2002 University of Alberta mathematician David Pimm wrote:

> Transformation is a key power of algebra, the most important means
> for gaining knowledge...In algebra, we can use symbols to express

a perceived relationship and then manipulate the result to obtain new information and insights. Indeed, these transformations embody the essential power of algebra because, due to their concise and non-physical nature, symbols are much more easily manipulable than the things they represent.[1]

This has certainly been on display throughout the book. One of the best examples of this occurred in Chapter 5, where we encountered word problems that were initially presented to us as a mix of pure English and verbalized numerical relationships. Our task there was to distill the quantitative essence out of such presentations.

Algebra provided the symbolic language for this process, which in that chapter I termed the "jump for joy stage"—not an official mathematical phrase, of course, but somewhat popular with many of my algebra students who were living the experience in real time. Once this stage is reached, a wide host of options became available to us.

The most important and dramatic part of it all is the entire track of transformation from the presentation of the word problem in original form to the "jump for joy stage" (where the equation describing the situation is obtained), on through the symbolic maneuvers that reduce the equation to its simplest form, yielding the numerical value for the unknown and then, where applicable, cascading it back to find other unknowns as well.

But there are also other interesting items of note along that track, if we care to take a gander.

One occurs during the simplification phase of the equation, the point just after we have obtained the equation in its raw form (jump for joy stage) and simplified it, yet before the point where we start to move objects from one side of the equation to the other.

Let's go back and revisit Word Problem 7 in Chapter 5, which states:

A total of $17,252 is divided up into one-, five-, and ten-dollar bills. If the number of fives is seven times the number of ones and the number of tens is quadruple the number of ones, how many of each type of bill are there?

For this problem, the regime of interest is from

$$1(x) + 5(7x) + 10(4x) = 17252$$

to

$$x + 35x + 40x = 17252$$

to

$$76x = 17252.$$

Now, instead of going on through and solving the equation, let's stop and ask the question: Can the simplified expression $76x$ on the left-hand side of the equation tell us anything in the context of the problem?

If you remember, in the original problem, there are three unknowns that harmonize according to the interrelationships specified—"the number of fives" is "seven times the number of ones" and "the number of tens" is "quadruple the number of ones." In a sense, these unknowns are really three moving targets that vary or dance in concert together, and along with them the total amount of money we have varies:

	Number of Bills	Amount of Money (in Dollars)
One-dollar bills	x	$1x$
Five-dollar bills	$7x$ (seven times the number of ones)	$5(7x) = 35x$
Ten-dollar bills	$4x$ (quadruple the number of ones)	$10(4x) = 40x$

Our ability to tag the three unknown dollar bill amounts as relationships in x demonstrates that for this problem each of the three unknowns varies in the same fundamental way. Or put algebraically, the variations that describe each amount are of the same fundamental type—each has a part that is some multiple of x (some number times x).

What the simplified expression essentially tells us is how all of that coordination and variation collectively mix together. That is, after combining the like terms together, the net result is equivalent to getting a bang of $76 in value times the number of one-dollar bills we have. This means that we can momentarily, if we want, switch how we look at the problem, viewing it now not as having three separate denominations of paper money but instead as a problem involving one new enhanced denomination that consists of hypothetical 76-dollar bills (whose internal composition breaks down according to the originally prescribed relationships between the three denominations). And now the question becomes, how many 76-dollar bills are needed to give a total of $17,252?

In this momentary viewpoint, it is important to note that the number of hypothetical 76-dollar bills matches the number of actual one-dollar bills. One of our 76-dollar bills contains 1 one-dollar bill plus 7 five-dollar bills and 4 ten-dollar bills. The following diagram illustrates the components:

If $x = 10$ (there are ten 76-dollar bills), then the total amount of money we have is $76(10) = 760. Now, if we wanted to see how this breaks down via the internal structure of the hypothetical enhanced bill, we simply multiply each of the constituents by ten. This yields 10 one-dollar bills, 70 five-dollar bills, and 40 ten-dollar bills for a total of $760. Another way to think about this is that if we had ten physical 76-dollar bills, we could take them to the bank and have the teller give us change in the following amounts as 10 ones, 70 fives, and 40 tens:

If $x = 75$ (there are seventy-five 76-dollar bills), then the total amount of money we have is $76(75) = \$5700$. Internally the money actually breaks up into the following amounts: 75 one-dollar bills, 525 five-dollar bills, and 300 ten-dollar bills:

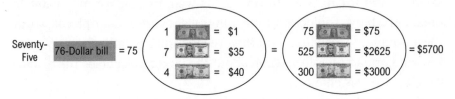

Try this interpretation out on the actual solution scenario for Word Problem 7 and see if you can organize the number of bills into 227 groups of "76-dollar bills," each consisting of 1 one-dollar bill, 7 five-dollar bills, and 4 ten-dollar bills according to the prescribed interrelationships.

This hypothetical viewpoint wasn't what we were particularly interested in at our initial reading of the problem, as we just wanted to use the expression to find out what value of x makes $76x$ yield $\$17{,}252$, and from this obtain the breakdown into the combination of one-, five-, and ten-dollar bills needed to satisfy the stated interrelationships. But this particular interpretation emerges from the algebraic maneuvers and is there for the taking if we so desire. It also yields a couple of observations that are worth mentioning.

INTERPRETATIONS OF THE MEANING OF AN EQUATION OR ITS SOLUTIONS

For the problem just discussed, a much clearer view of what is happening was obtained by coalescing together into one combined object the individual pieces of information related to the three denominations of money. Moreover, that combined object ($76x$) had an interesting interpretation all on its own.

Once the jump for joy stage was reached, we could have, before combining the values, tried to find a value of x all at once such that $x + 35x + 40x$ added up as three individual components to equal $17{,}252$. But that would be a harder thing to do versus first forming the combined object to yield the singular $76x$ and then dividing $17{,}252$ by 76 to obtain 227.

Afterward, it is a simple matter to go back and find out how many of each of the constituent denominations there are.

This notion of combining a bunch of interacting objects into a new transformed object is very much present in higher mathematics and science. And it can work in reverse as well, in that we may want to break up a bigger object into meaningful subcomponent pieces, study these smaller pieces first, and then reconstitute what is found to gain new and impactful information about the larger object (see the discussion later in this chapter).

This type of interpretation can also be used in a similar manner on the combined objects in the other word problems in Chapter 5. For those problems, however, there is a slight difference in that they consist of two fundamental components instead of one component, as demonstrated by the single term $76x$ for the bills problem. In looking at those other problems, it can be observed that adding up all of the relationships for the various unknown parts of information in each respective problem ultimately simplifies to an expression of the form (some number) times x + (a constant value).

This can be seen with Word Problem 3 in Chapter 5:

> A 570-foot rope is cut into three pieces. The second piece is four times as long as the first, and the third piece is ten feet longer than twice the first piece. How long are the three pieces?

The regime of interest for this problem is

$$x + 4x + 10 + 2x = 570$$

to

$$7x + 10 = 570.$$

Before going any further, let's ask what types of interpretation can be given to the simplified expression $7x + 10$ on the left-hand side of the equation. We will give a couple.

One way to think of this is to imagine the full 570-foot-long uncut rope that we want to subdivide according to the stated relationships between the three pieces. We can guess a value for the length of the first piece, mark it, and then measure out seven such lengths (including the first piece). We then see if the remaining length of rope after the last mark is 10 feet, to yield the entire 570 feet of rope.

Of course, going ahead and solving the equation will give us the desired value of 80 feet in length ($x = 80$). Once known, we can measure out 80 feet of rope and cut it; this gives the first piece. Then, we can measure out 4 times 80 or 320 more feet of rope and cut it to get the second piece. This would then leave the third piece, which should measure $2 \times 80 = 160$ feet plus the additional 10 feet to give 170 feet, which of course matches our earlier determination.

Another, more radical way to interpret the situation—described by $7x + 10 = 570$—is to look at the three individual unknowns in the problem as combining to give a collective bang of 7 feet. If the rope lengths were money as in the last problem, the equivalent would be to think of each of these 7-foot lengths as a seven-dollar bill. In order to include all of the information in the problem, however, we must hold back in reserve 10 feet of rope to be added to the third piece at the end.

In this viewpoint, we can then ask, how many 7-foot-long rope segments must we have such that at the end, when we add the 10-foot segment of rope held in reserve, we have a combined total of 570 feet of rope? Solving this equation tells us that we will have 80 such segments—which gives 560 feet—plus the 10 feet of rope.

Once we find the number of rope segments, we can imagine taking each 7-foot segment and breaking it apart according to the initially given interrelationships between the pieces: the second piece is four times as long as the first and the third piece is twice the first. This will yield a breakdown for a given 7-foot segment into the first piece being 1 foot long, the second piece being 4 feet long, and the third piece being 2 feet long:

| 1 ft | 4 ft | 2 ft |

[Image provided courtesy of William Hatch]

Continuing on with this hypothetical situation, we can now reimagine the three unknowns as being represented by three containers. We cut and collect in the container for the first piece 1 foot of rope from each of the eighty 7-foot segments. This yields 80 individual 1-foot-long segments of rope that sum to a collective total of 80 feet. We cut and collect in the second container the 4-foot segments from each of the 80 segments to yield a total of 320 feet of rope. In the third container, we collect the remaining 2 feet of rope from each of the 80 segments to obtain 160 collective feet of rope, to which we also add the additional 10 feet of rope held initially in reserve to give a final value of 170 feet of rope. Check that the total amount of rope assigned for each of the three containers given here matches the correct values for the three pieces discussed earlier and in Chapter 5.

This time, however, unlike the first interpretation presented, there is a big difference between the problem as initially stated and this second hypothetical interpretation. In the former, three continuous pieces of rope that satisfied certain specified conditions were asked for and then discovered to be of lengths 80 feet, 320 feet, and 170 feet. But now, in this second interpretation, we are dissecting the rope into a great many segments and then aiming to find the sum total length of rope that we must place into each of three containers. This construction ends up yielding 80 individual 1-foot-long segments of rope in the first container, 80 individual 4-foot-long segments of rope in the second container, and 81 segments in the third container (80 individual 2-foot-long segments plus 1 segment 10 feet long). This is quite a different physical situation.

Thinking of the problem in either of these vastly different physical interpretations presents no issue, as long as all we ultimately want is to use algebra as an aid in solving the problem as originally stated. It would be quite a different matter, on the other hand, if we were trying to instruct someone on how to physically perform the desired construction.

One may skeptically question the realism of either of these problems. But as mentioned in Chapter 6, whether or not the problems themselves are realistic is not as important as the educational illumination that they can sometimes cast on the relationships and situations that do occur with great force and relevance elsewhere.

Science and engineering offer many examples where symbolically maneuvering equations, expressions, or data or changing the viewpoint or interpretation of their solutions (or forms) can yield pay dirt. One of the most dramatic examples of this occurred in the 1860s in the field of electromagnetism. The star theoretician James Clerk Maxwell pieced together and wrote the foundational narrative of the subject by synthesizing together a group of principles and equations much of which had been discovered and articulated earlier in the century, in a piecemeal fashion by many prominent figures (some of whom were mentioned late in Chapter 10). To this he added his own essential contributions and extraordinary insight, an insight often compared with those of Newton and Einstein themselves.

Maxwell then manipulated the equations, looking at them from various viewpoints, and in the process discovered that they predicted the existence of hypothetical electromagnetic waves. From this, he proceeded to compute the speed of these mathematical waves and found that their predicted speed matched that of the known speed of light, which had been calculated through other independent means starting with Danish astronomer Ole Rømer's successful approximation in 1676 (and known now to be about 670+ million miles per hour or 186,000+ miles per second).

Maxwell realized that the shock of this coincidence was too much to be accidental and hypothesized that visible light itself must be a particular version of these electromagnetic waves. Moreover, his analysis revealed that there should be other electromagnetic waves with different wavelengths than that of light.

At the time, through earlier experiments that had extended information to just beyond both ends of the visible light spectrum (red and violet), infrared light and ultraviolet light had already been discovered. Maxwell hypothesized that these too must be electromagnetic radiation, and furthermore predicted the existence of other electromagnetic waves. These further predictions were confirmed in the late 1880s, 1895, and 1900 with the discovery of radio waves, X-rays, and gamma rays, respectively.

So, though the simple re-interpretations we engaged in with our simplistic word problems here are quite trivial in terms of direct meaning, imagine for a moment if they were actually different viewpoints to some scientific analysis, and that the alternate way of looking at the money problem—as 76-dollar bills—or the rope problem—as 80 segmented pieces—opened up a whole new way to look at some phenomenon; and you will get a small taste of what can and sometimes does occur in science and mathematics.

It goes without saying that Maxwell's interpretations regarding the existence of electromagnetic waves is about as revolutionary an insight as you will find in any science, and the subsequent applications of such waves—radio, television, X-rays in medicine, and microwave ovens—have literally transformed our world.

CLOSURE

Let's now look a bit closer at the fact that the word problems in Chapter 5 all simplified to expressions of the same form: namely, (a number) times x + (a constant value).

If we look at each of these simplified problems—from jump for joy stage to simplified form—as a different scenario, we can use parameters to represent them all by expressions of the form $Ax + B$. (We use uppercase letters here to distinguish these "dual-acting parameters," as will become clear later.) For the time being, we are only looking at the left-hand side of those equations.

Thus, for the rope lengths situation we would have $A = 7$ and $B = 10$, which would give the form $7x + 10$ on the left-hand side; and for the money situation we would have $A = 76$ and $B = 0$, yielding $76x + 0$ or simply $76x$. For Word Problem 4 from Chapter 5, we have that $A = 5$ and $B = 20$, and for Word Problem 5 we would have $A = 8$ and $B = -302$, corresponding to $5x + 20$ and $8x - 302$, respectively. All five of the problems in Chapter 7 simplify to the situation $16x + 10$, which corresponds to $A = 16$ and $B = 10$.

The fact that the individual pieces of information in the various word problems all combine to an expression of the same form is representative of an important mathematical property called *closure* under an operation. Before discussing this a little more in the context of the word problems, let's look at closure in a few other scenarios:

1. The even numbers are closed under the operation of addition (under addition): Two even numbers added together always yield another even number.
2. The odd numbers are not closed under addition: Two odd numbers added together don't yield another odd number but an even number.
3. The integers $(\ldots, -3, -2, -1, 0, 1, 2, 3, \ldots)$ are closed under both addition and multiplication:
   - Two integers added together always yield another integer.
   - Two integers multiplied together always yield another integer.
4. The nonzero integers $(\ldots, -3, -2, -1, 1, 2, 3, \ldots)$ are not closed under division: Dividing one integer by another integer does not always yield another integer. For example:

   - $\dfrac{-400}{25} = -16$ is an integer.

   - $\dfrac{9}{5} = 1.8$ is not an integer.

   - $\dfrac{7}{20} = 0.35$ is not an integer.

5. The common fractions (or rational numbers), which can be written as $\dfrac{\text{one integer}}{\text{another integer}}$ (or using alphabetic symbols as $\dfrac{x}{y}$, where both x and y are integers with $y \neq 0$), are closed under addition, subtraction, multiplication, and division (where defined): Two rational numbers added to, subtracted from, multiplied by, or divided by one another (where defined) always yields another rational number. For example:

   - $\dfrac{1}{5} + \dfrac{2}{7} = \dfrac{7}{35} + \dfrac{10}{35} = \dfrac{17}{35}$.

   - $\left(\dfrac{3}{4}\right)\left(\dfrac{5}{11}\right) = \dfrac{15}{44}$.

6. Expressions of the form $Ax + B$ (where A and B are real numbers) are closed under addition.
   - Take two such expressions $5x + 70$ and $11x + 20$, and add them together to obtain $(5x + 70) + (11x + 20) = (5 + 11)x + 70 + 20 =$

16x + 90, which is still of the same form $Ax + B$ with $A = 16$ and $B = 90$.

- ○ This holds in general. Take two general expressions $ax + b$ and $cx + d$, where a, b, c, and d are real numbers. Adding them together yields $ax + b + cx + d$, which can be simplified to $(a + c)$ $x + b + d$. This is still of the same form $Ax + B$ with $A = a + c$ and $B = b + d$. In the previous example, we have $a = 5$, $b = 70$, $c = 11$, and $d = 20$.

What, if anything, is to be gained by knowing that expressions that can be written in the form of $Ax + B$ are closed under addition? Two things immediately come to mind.

Firstly, a general method of solution can be implemented for all of the word problems posed in Chapter 5, as well as thousands more similarly posed problems. That is, if you know how to solve equations such as $76x = 17292$ or $7x + 10 = 570$, then the general equation $Ax + B = F$ can be solved in a similar manner. We have introduced the parameter F on the right-hand side of the equation to complete the picture. In the two equations just shown, F is equal to 17,292 and 570, respectively.

The general solution to this equation can be found as follows:

Original Equation   Equation in Simplest Form

$Ax + B = F$ — Subtract B from both sides → $Ax = F - B$ — Divide both sides by A → $x = \dfrac{F - B}{A}$

Once we find x, we can of course cascade it to find other unknown pieces of information according to the relationships specified in each particular problem. For instance, for Word Problem 4 in Chapter 5, the simplified form is $5x + 20 = 180$. This gives $A = 5$, $B = 20$, and $F = 180$. Thus, putting these values into the simplest form for x in the reduction diagram gives

$$x = \frac{180 - 20}{5} = \frac{160}{5} = 32.$$

From this we can cascade it according to the given relationships to find the two other angle values.

This general form can now be used in any similar type of problem whatsoever; thus, in one grand gesture we can represent the solutions to all of these problems by this all-encompassing symbolic maneuver.

We see this in the brief description of quadratic equations in Appendix 1, where it is mentioned that all quadratic equations in any form are ultimately of the same type. They can be described by the one super-equation $ax^2 + bx + c = 0$ and then solved by the same technique, which ultimately yields the quadratic formula.

The second thing that comes to mind is that similar types of interpretations as we have given for the money and rope piece problems can also be given to problems expressible in the $Ax + B$ form.

Let's pose a more general version of Word Problem 7 involving the money denominations discussed previously, complete with parameters: A total of F dollars is divided up into one-, five-, and ten-dollar bills. If the number of fives is a times the number of ones and the number of tens is b times the number of ones, how many of each type of bill are there?

We can set this problem up as before with the following breakdown:

	Number of Bills	Amount of Money (in Dollars)
One-dollar bills	x	$1x$
Five-dollar bills	ax (a times the number of ones)	$5(ax) = 5ax$
Ten-dollar bills	bx (b times the number of ones)	$10(bx) = 10bx$

We know that adding up the money from all three bill types must yield the total amount of money (now given by F dollars). Comparing side by side the progression from the jump for joy stage to the simplified form for both problems yields the following:

Steps in Original Concrete Word Problem	Steps in General Word Problem
$1(x) + 5(7x) + 10(4x) = 17252$	$x + 5(ax) + 10(bx) = F$
$x + 35x + 40x = 17252$	$x + 5ax + 10bx = F$
$(1 + 35 + 40)x = 17252$	$(1 + 5a + 10b)x = F$
$76x = 17252$	$Ax = F$

Note that for the form $Ax + B = F$, we have here that $A = (1 + 5a + 10b)$ and $B = 0$.

Using a similar interpretation as we did for the hypothetical 76-dollar bills, we now have a hypothetical $(1 + 5a + 10b)$-dollar bill that breaks down as follows:

Note that for $a = 7$ and $b = 4$, this becomes a 76-dollar bill, and this diagram becomes equivalent to the original diagram we presented. However, this new diagram is far more general. So, for instance, if we let $a = 9$ and $b = 11$, we get a hypothetical 156-dollar bill that breaks up into 1 one-dollar bill, 9 five-dollar bills, and 11 ten-dollar bills.

We are not completely free to choose the parameters at will in this money scenario, however, as they have to all be positive values and we additionally have to make sure that $(1 + 5a + 10b)$ divides evenly into F. For example, $F = 1248$ will work in the 156-dollar bill case, yielding 8 one-dollar bills, 72 five-dollar bills, and 88 ten-dollar bills. This requirement is based on the fact that the problem has to both be logically consistent and describe possible situations regarding money in US currency—we can't have a negative number of bills or a bill denomination that corresponds to one-fifth of a dollar. But that still leaves quite a bit of variety in the types of word problems we can analyze—differing in the details but similar in type and interpretation.

Similar setups could be done for the other word problems in Chapter 5, with additional modifications for when B is nonzero. We are assured of this because of the fact that we know all such problems are of the same type and can ultimately be simplified to and represented by the same type of equation (the closure property): $Ax + B = F$. Closure allows us to corral them all under a single tent.

Closure under an operation turns out to be a crucial component in many other more advanced kinds of algebra, along with other types of important properties. These systems are part of one of the most important regions of mathematics, abstract algebra, which includes such fields as group theory, ring theory, field theory, module theory, representation theory, and linear algebra. Astonishing applications both within and outside of mathematics have been uncovered for some of these areas, particularly group theory, representation theory, and linear algebra. Such applications have occurred in a wide variety of places, including geometry, topology, quantum physics, particle physics, relativity, solid state science, chemistry, some forms of spectroscopy, data science, internet search algorithms, cryptography, error correcting codes, combinatorics, and so on.

A mighty river indeed this abstract algebra is, with many powerful tributaries feeding it, yet the symbolic innovations in the development of basic algebra occurring in the sixteenth and seventeenth centuries remain a primary direct source. The simplest renditions in this groundbreaking achievement live on in perpetuity in the schoolroom algebra of today.

QUANTITATIVE COCKTAILS AND ATOMIC SPECTRA

Let's reverse gears and consider situations where complex encoded information can be broken down into meaningful subcomponents for great gain. It is an especially general and cross-disciplinary idea with spectacular materializations across the breadth of human experience. We begin with the highly interdisciplinary subject of spectral analysis.

In the late 1850s, famed German chemist Robert Bunsen—of Bunsen burner fame—wrote to a friend:

> At the moment I am occupied by an investigation with Kirchhoff which does not allow us to sleep. Kirchhoff has made a totally unexpected discovery, inasmuch as he has found out the cause for the dark lines in the solar spectrum and can produce these lines artificially intensified both in the solar spectrum and in the continuous spectrum of a flame, their position being identical with that of Fraunhofer's lines. Hence the path is opened for the determination of the chemical composition of the sun and the fixed stars with the same certainty that we can detect [strontium chloride], etc., by our ordinary reagents. By this method the chemical elements occurring upon the earth may also be detected and separated with the same degree of accuracy as upon the sun...[2]

A landmark moment in the history of astronomy—indeed all of science—that opened to the world in a truly systematic way the vast field of spectroscopy. Astronomer Carl Sagan described it thus in 1980: "Astronomical spectroscopy is an almost magical technique. It amazes me still."[3]

What is it about the subject that so excited Bunsen and Sagan in their respective eras as well as legions of other scientists up to the present? Undoubtedly, much owes to the subject's clever ability par excellence to extract detailed information in signals from unbelievably remote and inaccessible sources. The manner in which it has been used to reveal chemical and physical knowledge of the heavens, as well as so much here on Earth, is reminiscent of the abilities of mathematics as a diagnostic instrument in analyzing a broad array of questions, with algebra and arithmetic being essential tools in these investigations.

The discipline stands up well against the very best that science has to offer, with nearly two dozen individuals having received Nobel Prizes for discoveries, revolutionary perceptions, and inventions that have been related to some aspect of spectroscopy.[4] Here, we try to give just a tad bit of insight into this capital and highly relevant field by coming at it tangentially through a few of the examples discussed in this text—using them as conceptual fuel.

In 1666, Isaac Newton opened the door to the world that spectroscopy would become by passing white light through his prism and uncovering the fact that it was not at all the single ingredient that it appears to be, but rather a diverse cocktail of the major colors of the rainbow. The prism naturally takes the combined object, white light, and splits it into its constituent colors or ingredients. Later it was revealed that these different ingredients can be distinguished and identified by their wavelengths, which determine the differences in how they pass through the prism. The following numerical scale shows the sizes of some of these wavelengths, which are extremely small, in Angstroms (1 nanometer = 10 Angstroms or approximately 0.00000000328 feet), abbreviated Å:

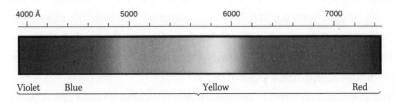

Wavelengths given in Angstroms [Image from *UT Austin–Principles of Chemistry* (Li et al. 2018)]

From this spectral reading, we can see that the colors visible to the human eye have wavelengths that range from violet (around 3900 Angstroms) to red (around 7400 Angstroms).

Can we do this with the quantitative cocktails we have discussed? That is, can we look at the blended information obtained there and definitely determine the exact contributions from each of the constituent ingredients that make it up?

Specifically, in the language of course averages, given an overall course average for a student of 80%, can we figure out the exact scores that this student made on all of the homework assignments, each of their tests, and the final exam? Is there a mathematical prism that will allow us to unambiguously break up that 80% score back into its component parts?

Our discussion in the first part of Chapter 10 bore critically on this question. There, we saw that, even if we dial it back from every

individual assignment to just the averages on each of the three categories of homework, tests, and the final exam—at contributions of 20%,
60%, and 20%, respectively—the answer to this question is still in the
negative. We found that the equation $0.20x + 0.60y + 0.20z = 80$ was indeterminate, and that there were many possible varying combinations
of scores that could all coalesce to yield the same course average of
exactly 80%. So, without more information, we can't recapture exactly
how the individual component scores combine to yield the 80% score.

What about the number of days and age problem and the data generated there?

Consider the situation with the same criteria as discussed at the
beginning of Chapter 8, with the date being June 30, 2021. For those
who satisfy the criteria of the problem, being younger than 100 years of
age and having had a birthday by this date, can their number (without
knowing the person) be parsed out into the two ingredients that combine to make up its value—namely, the number of days they like to eat
out in a week and their year of birth?

The answer this time is in the affirmative, due in part to the fact that
though these ingredients combine to form a three-digit number, they
don't interfere with each other in a way that loses the information. They
combine in a block-like fashion such that their tracks in the formation
of the final number can be uncovered when algebra is employed.

For example, if in 2021 someone running through the procedure generates the number 647, it is straightforward to show how this breaks
down into the two unknown components that make it up. We saw in
Chapter 2 that the algebraic representation of the entire procedure reduces to the expression $100x + (2013 + z - y)$. For the year 2021, z is 8, so
the expression becomes

$$100x + \underbrace{(2021 - y)}_{\text{age}}.$$

Here, x represents the number of days the person likes to eat out in a
week and y represents the calendar year the person was born.

Our knowledge of how this problem works means we can partition
647 as $600 + 47$, and then solve the following two equations (often in our

head): $100x = 600$ and $2021 - y = 47$. Solving both gives us that the person likes to eat out six days in a week and that their year of birth was 1974.

The recovery succeeds here because the relevant items, 600 and 47, are retained whether we write them separately or add them to obtain 647. In general, for a given year, we can break down all such magical three-digit numbers this way as long as the person is younger than 100 years old and we know whether or not they have had a birthday.

For a person who is 100 years or older, we have a three-digit number for the age that will interfere with the digit in the hundreds place. For example, for a person 105 years old, we would have $600 + 105$, which becomes 705. This number fails to properly decompose using the simple solving procedure for this problem.

HOT GAS CLOUDS

Amazingly, in the case of light from distant sources in the universe, the situation is more analogous to resolving the magical three-digit numbers back into their component unknowns. The light, which in some cases comes from trillions of miles away, contains data that is encoded in such a way—as a cocktail of rays of different wavelengths—that it can yield useful diagnostic results. Often the components that make up this complex mix of information retain enough of their individuality to allow them to be selected back out using the right equipment and analysis. This often yields valuable information about both the nature of the cosmic sources that produced the light and the physical matter with which the light has interacted along its journey to Earth.

Visible light can be produced from chemical elements whose atoms have been heated up or excited, but they don't produce a continuous rainbow-like spectrum like the white light that Newton observed; instead, they produce a spectrum that shows individual lines of varying colors. This line spectrum turns out to be unique for each particular element. The following diagram displays the visible emission spectra (with the more prominent lines) of the first 99 elements of the periodic table:

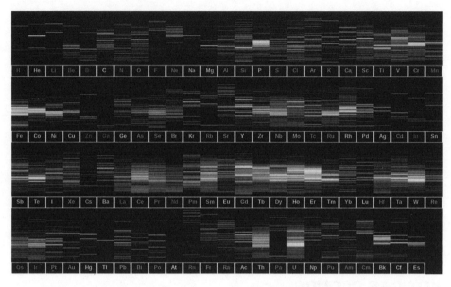

Spectral lines for the elements (presented vertically) in black and white; the shorter wavelengths (blue) are at the bottom for each element's spectrum, and the longer wavelengths (red) are at the top [Reprinted with kind permission from Julie Gagnon at umop.net/spectra]

The line configurations of each of the elements are distinct from each other, meaning that such patterns can work like fingerprints to identify the presence of particular elements in the sources that produced the light. Scientists labored hard in the time following Kirchhoff's and Bunsen's discovery to generate the spectra of the many elements. In fact, such methods even allowed for the discovery of new elements.

Here are the prominent visible emission line spectra of the elements sodium (Na), hydrogen (H), calcium (Ca), and mercury (Hg) for comparison.

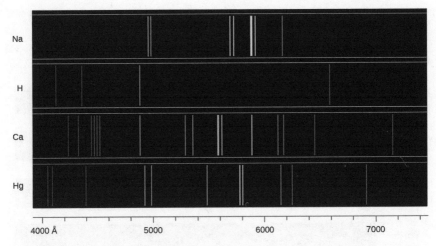

Line spectra of the visible light from excited sodium, hydrogen, calcium, and mercury atoms (presented horizontally) [Image from *UT Austin–Principles of Chemistry* (Li et al. 2018)]

Thus, when light from a source such as a hot gas in a nebula reaches Earth as a cocktail of light rays of different wavelengths, we can be fairly certain that these wavelengths have been produced or affected by the various elements present in the gas. The question is, can we tell which elements?

The way to separate out the light cocktail into the characteristic light rays of the elements present is through the use of specialized instruments, such as a spectroscope, which employ devices that are far more sophisticated at dispersing a light beam into its component beams than the simple prisms such as the one that Newton used. The line spectrum so generated may be a mix of many spectral lines. If an astronomer notices in the pattern prominent lines characteristic of a particular element, such as the distinct double lines just less than 6000 Å (and bright yellow in color) characteristic of sodium or the quartet of lines just above 4400 Å (and blue in color) characteristic of calcium, they can feel fairly justified in the conclusion that excited sodium or calcium is present in the gas cloud.

The next diagram gives an example of what an observed spectrum may look like. Below it are some of the brighter lines from the characteristic spectra of five important elements:

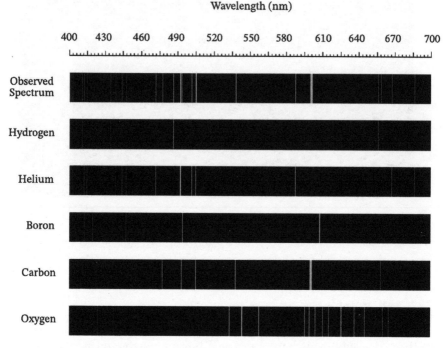

Observed spectrum (wavelengths given in nanometers) and spectra of some potential component elements [Image provided courtesy of Nagwa Limited]

(Recall that 1 nanometer = 10 Angstroms or approximately 0.00000000328 feet; thus, 400 nm = 4000 Angstroms, 700 nm = 7000 Angstroms, and so on.) In this particular observed emission line spectrum, many lines matching those in hydrogen, helium, and carbon are present. The lines for oxygen and boron don't match lines in the observed spectrum. Based on this, a reasonable conclusion would be that hydrogen, helium, and carbon are producing this observed light coming from the nebula, whereas oxygen and boron are not.

To give another view of what is happening, the next diagram provides a visual of a hypothetical observed spectrum from an imaginary nebula and also hypothetical elements whose lines are distinguished by shapes rather than colors. In the top diagram, notice that lines matching the spectra for Elements A, C, and E are present in the observed spectrum. We can conclude from this matching that these elements are most likely present in the nebula.

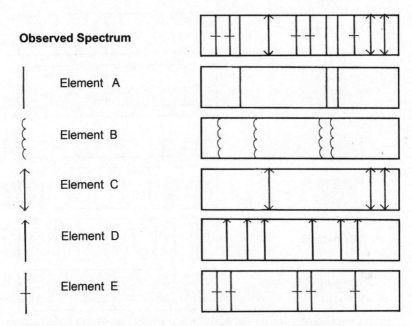

Hypothetical line spectrum where Elements A, C, and E are present in the observed spectrum [Artwork provided courtesy of William Hatch]

In the following diagram, we can see lines in the observed spectrum matching lines from Elements B and D, meaning those elements and not Elements A, C, and E are likely producing the observed light in this case.

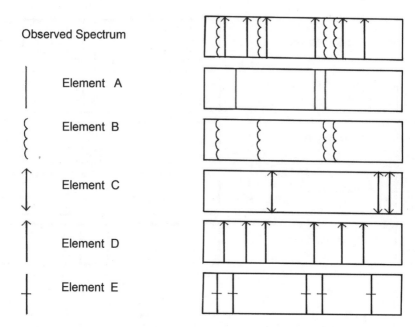

Hypothetical line spectrum where Elements B and D are present in the
observed spectrum [Artwork provided courtesy of William Hatch]

ABSORPTION SPECTRA: STARLIGHT

Light from stars, including the sun, share spectral information in a different way than do the emission nebulae just discussed. Their spectra are not bright lines but rather a continuous rainbow band like the one that Newton observed when performing his prism experiments on sunlight. However, on closer examination through a spectroscope, we find that such spectra are not completely continuous, but rather can show certain dark lines peppered throughout the rainbow band of colors. These dark lines are called absorption lines and form what is called the absorption spectrum of the star. Just like the bright emission lines of nebulae, they too contain information.

Stars consist of a hot, dense interior of gases surrounded by a layer of gases called the photosphere (often called the surface of the star). Many stars, like the sun, also have additional atmospheric layers called the chromosphere and the corona. It is the light from the photosphere that we observe when we see the sun and other stars. By the time a particle of light or photon reaches and exits the photosphere, it has traveled a long way from the deep stellar interior where the nuclear fusion that produced it occurs.

By this time, such a photon may have been involved in billions of interactions, continually being absorbed and re-emitted by atoms in the interior. In reality, the photon at the surface represents the last generation of a cascade of energy-matter interchanges. It is estimated that this process can take tens of thousands of years and these continual interactions have an averaging effect on the energy of the photons that reach the photosphere and then travel to Earth, meaning that a large number of the wavelengths of visible light in the range from 3900 to 7400 Angstroms (390 to 740 nanometers) are present in the observed spectrum. This accounts for the continuous-appearing rainbow band spectrum observed in Newton's prism experiment, which was not sensitive enough to detect the dark lines.

The dark lines in the spectrum come from elements that are present in the photosphere. In experiments performed on Earth, it was observed that elements can absorb light of the identical wavelengths in which they emit light. In terms of quantum theory, this corresponds to an electron in an atom going from a lower energy level to a higher energy level: Emission of light occurs when it goes from a higher energy level to a lower energy level, but the energy difference is the same either way. This means that, on Earth, if you placed relatively cooler sodium vapor in between white light and a spectroscope, the sodium atoms would absorb light of the same wavelength that they emit light in the excited state, and the observed spectrum would be a continuous rainbow band peppered by dark lines that would be in the exact locations—by wavelength—that the bright lines appear in the emission spectrum of sodium. This was one of the big findings in Kirchhoff's and Bunsen's great epiphany regarding the solar spectrum, whose dark bands were first observed by William Wollaston in 1802 and more intensely analyzed and reported on by Joseph von Fraunhofer beginning in 1814.

Another way to think of it is that just like a tree can block or absorb sunlight to create a shadow, so too could gas surrounding a star that contained sodium vapor absorb or block out light. The difference here is that the tree blocks out all of the visible light when it creates a shadow, whereas in the case of sodium, the vapor acts as a filter due to absorption, and only blocks out a certain type of light—exactly the same type of light that excited sodium emits. So, the observed spectrum of the star contains sodium shadows.

In the following diagram of the sun's spectrum, the shadow of the prominent sodium lines can be clearly seen at position D at approximately 589.0 and 589.6 nanometers (5890 and 5896 Angstroms). The C line corresponds to a hydrogen shadow at 656.3 nanometers (6563 Angstroms). The A and B lines correspond to oxygen shadows. Many of the other shadow lines in this spectrum correspond to lines from iron.

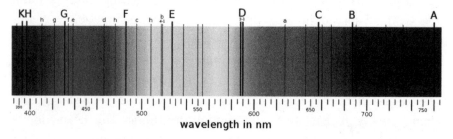

The sun's spectrum [Image is in the public domain]

From these readings, astronomers have been able to conclude that there is iron, sodium, oxygen, and, of course, hydrogen in the photosphere of the sun. The same techniques can be used to identify elements that are present in more distant stars and other astronomical objects. These days, however, ever more sophisticated instruments and computer software are employed to greatly enhance the analysis, even allowing information from the atmospheres of planets around other stars (exoplanets) to be obtained through their atmospheric absorption of the light from the primary star that they orbit.

Moreover, what can be done with atoms can also be done with molecules (chemical combinations of two or more atoms). The emission and absorption processes in molecules are different and more varied than they are for individual atoms. This consequently translates to different types of spectra, often outside the visible wavelength range, but the general idea of identification of substances remains and can be extremely useful both astronomically as well as here on Earth.

The visible spectrum is only one portion of the vast electromagnetic spectrum. Observations made on these other types of non-visible electromagnetic radiation—including radio wave, infrared, ultraviolet, and X-ray wavelengths—can tell us even more about the composition and activities of many objects in the cosmos. Using the results from spectral analysis, astronomers have come up with entire classification schemes for stars, nebulae, and other astronomical objects. Spectral characteristics can even be used to reveal the temperatures of astronomical objects and whether or not they are moving toward us or away from us. These are not small things, as most such objects in the nighttime sky are trillions of miles away from us, and for astronomers to be able to identify many of the substances that make them up, as well as other key indicators from the coded information in light, remains one of the landmark collective achievements of science. It truly is the gift that simply keeps on giving.

A GRAND IDEA

As fundamental and groundbreaking as spectral studies in the heavens have proved, they are just one of the most visible manifestations of a powerful idea that can be found throughout mathematics, science, and engineering (see this chapter's epigraph by Ingrid Daubechies).

Hearkening back to our early discussions in Chapter 4, we mentioned that one could learn basic things about arithmetic just from playing with pebbles in the dirt of our backyard. We further likened the idea of learning some arithmetic this way, to the materialization—in our backyard—of the spirit of the vast and broad subject of arithmetic. Similarly, though we have discussed spectral analysis in a specific and foundational context—"our backyard"—it reveals itself in other areas such as theoretical quantum mechanics, nuclear physics, radar, sonar, and signal processing in general.

For instance, with sonar, sound waves from various activities in the ocean are received by a detector on a submarine, ship, or underwater device. The task of sonar technicians is then to take such signals and try to extract information from them by decomposing the received sound into subcomponent pieces for analysis. The idea rests in the knowledge that different objects should give off different sounds or acoustical fingerprints. A humpback whale makes a sound that is different from the sound made by a fin whale, which is different from the sound made by an enemy submarine (or surface ship going through the water), which is different from the sounds made by a large iceberg running aground or a hydrothermal vent.

Like astronomers who have in their possession the characteristic spectral lines for the many elements, the sonar technicians undoubtedly have a database of various sound signatures for different objects. Their desire is to break down the observed acoustical spectrum into information that can help identify what objects or entities are producing the sounds received by the detector. However, sound waves are not electromagnetic waves, and the breakdown of information from the audio spectrum is not as clean and definitive as the chemical information that is obtained from the electromagnetic spectrum. Consequently, mathematics and computer science weigh heavily in this diagnostic analysis, which falls under the important electrical engineering field of signal processing.

Another type of sonar called active sonar can involve a ship or submarine sending out its own sound signal as a probe and then analyzing the reflected signal on return. Such sonar allows technicians to obtain more detailed information—such as the distance to the object or its speed—but it also has the side effect of giving away the vessel's position to a listening enemy ship or submarine. This obviously can be highly disadvantageous in the case of military actions, where the need for stealth is at a premium. But there are many seagoing operations that are of a scientific or exploratory nature where the need for secrecy is not so important—mapping the ocean floor, locating navigational hazards, finding missing aircraft, ships, or submarines, or tracking underwater geological activity; and active sonar can be very useful in such cases.

Radar, used to detect airplanes or measure weather-related information in the atmosphere, works on a similar idea to sonar except that its devices detect and emit electromagnetic waves—radio waves and microwaves—instead of sound waves. Radio and microwaves have wavelengths much longer than the wavelengths of visible light. These wavelengths range roughly from 10 million to 1 trillion Angstroms (1 million to 100 million nanometers) depending on the specific type of radar used.[5] Radar can be used in an active or passive manner as well. Active radar is the more familiar of the two.

CONCLUSION

Famed linguist Noam Chomsky stated in 1970:

> Language is a process of free creation; its laws and principles are fixed, but the manner in which the principles of generation are used is free and infinitely varied. Even the interpretation and use of words involves a process of free creation.[6]

The same can be said of mathematics, science, and engineering. There is a great deal of freedom to roam in all of these areas, yet the logic of mathematics and the known laws of nature are always there to be reckoned with and respected.

Hopefully, this chapter has hinted just a bit at the additional treasure that can sometimes be gleaned in looking at mathematical and scientific results from different yet complementary viewpoints. In the first half of the chapter, we gave a different look to a word problem from Chapter 5, which then served as a gateway to briefly touch on the topic of closure under an operation.[7]

From using closure, we saw that it became possible to generalize both the method of solution as well as the new interpretation we gave to the problem, as involving a hypothetical larger bill denomination. Parameters continued to play a key role in allowing for the generalization of these efforts.

One may well ask, did these additional interpretations really assist us in better solving these word problems? Probably not, as the solutions are more efficiently obtained through the processes already discussed in Chapter 5. What these interpretations granted us was the ability to shed a tad bit of illumination on something that can often occur to far greater effect elsewhere in mathematics and science. That is, by simply reanalyzing a problem from a different viewpoint, whole new vistas may open wide to us—vistas that are so deep and vast, in some cases, that they can lead to groundbreaking insights into mathematics or into physical phenomena.

A case in point is the ancient Indians' decisive reimagining of numeration away from the much more common additive systems, such as the Egyptian hieroglyphic system or Roman numerals, to far more potent and extendable systems where positional values play a featured role. It is no stretch to say that this switch to the Hindu-Arabic numerals is one of the most important changes in operational viewpoints in the history of human thought. An in-depth discussion of the environment of this reorganization can be found in my earlier book, *How Math Works: A Guide to Grade School Arithmetic for Parents and Teachers*.

The second half of the chapter saw us continue in this vein by looking at course averages and the number of days and age problem from slightly different points of view, now seeing if the consolidated information they contained could be unambiguously unpacked into the basic elements involved in their construction. We found that we couldn't definitively do this in the case of course averages, but in the number of days and age problem it was indeed possible.

This latter unpacking was then used in a tangential way to mildly season the truly meaty subject of spectroscopy, where such unpacking is elevated to the level of one of the great wonders of science. We further saw that the grand idea that underwrites spectroscopy has other materializations in the analysis of phenomena much closer to home.

In the case of astronomical spectroscopy, the physical world has been extremely generous in the tools it naturally grants astronomers in the identification of substances. This allowed the field to progress far even before the advent of the computer. However, in the quest to extract more of the information contained in the electromagnetic signals received from space, this identification becomes a far messier operational challenge. This is also true for the operational challenges in other areas such as the acoustic signals received in sonar, the radio and microwave signals received in radar, the seismic signals received in geology, and the diverse electrical signals needed in a host of engineering applications, including communications and image processing. Here, sophisticated mathematical tools combined with high-powered computers to implement fast algorithms become critical features in handling many of these challenges.

In the early 1800s, a French mathematician—and one of Napoleon's scientific advisors and administrators—named Joseph Fourier made a remarkable discovery. He realized that you could take certain primary objects and break them up into radically new subcomponents that individually were totally unlike the main object. Yet, if you took enough of these new subcomponents together, their collective mass could behave like the big object, at least over a certain region. This decomposition was anything but natural. Yet it has proven to be extraordinarily powerful, with a reach that extends all of the way up to our current time. Along with its subsequent developments, it forms one of the most important tools in the analysis of all sorts of signals, including the sound signals in sonar and the radar and microwave signals in radar.

Imagine if the methods employed in the simplistic yet radical reinterpretation of the rope problem—involving the dissection of the 570-foot-long rope into 80 individual 7-foot segments—ultimately turned out more than 200 years later to be useful in a wide array of physical real-world problems—problems and challenges that were not even a glint in the eye of the original dissecting mathematician, at the time of the genesis conception of the idea—and you will get perhaps a glimpse of the miracle spawned by Fourier's original great analysis.

Motions of the Heart

To be emotionally stirred is to care, to be concerned.
It is to be *in* a scene or subject not outside of
it…The more anything, whether an object, scene, idea,
fact or study, cuts into and across our experience the more
it stirs and arouses…

No amount of possession of facts and ideas, no matter
how accurate the facts in themselves and no matter
what the sweep of the ideas—no one of these in
themselves secure culture. They have to take effect in
modifying emotional susceptibility and response before
they constitute cultivation.

—John Dewey (1859–1952), Address to Harvard Teacher's
Association, March 21, 1931

12

Grand Confluences

We must look to our own faculty for discerning those fine connective things—community of aim, interformal analogies, structural similitudes—that bind all the great forms of human activity and aspiration—natural science, theology, philosophy, jurisprudence, religion, art and mathematics—into one grand enterprise of the human spirit.

—Cassius Jackson Keyser (1862–1947), "The Humanization of the Teaching of Mathematics," *Science*

Our brains are complicated devices, with many specialized modules working behind the scenes to give us an integrated understanding of the world. Mathematical concepts are abstract, so it ends up that there are many different ways they can sit in our brains. A given mathematical concept might be primarily a symbolic equation, a picture, a rhythmic pattern, a short movie—or best of all, an integrated combination of several different representations...

A whole-mind approach to mathematical thinking is vastly more effective than the common approach that manipulates only symbols.

—William P. Thurston (1946–2012), Foreword to *Crocheting Adventures with Hyperbolic Planes* and Foreword to *The Best Writing on Mathematics 2010*

"Every thing throws light upon every thing," observed the "Yale Report of 1828"—the nineteenth-century's strongest bulwark for retaining what has been called the classical curriculum prescribed for all students attending college.[1] As discussed in Chapter 7, this curriculum was based on an extremely rugged ancient Greek and Roman core, and had come under increasingly intense criticism throughout the 1820s as being completely out of touch with student needs of the day.[2]

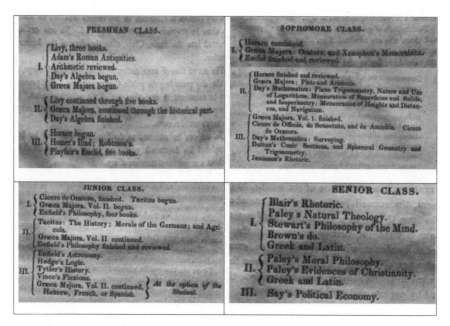

1828 Yale University Catalog Course of Instruction (pp. 24–25): "Livy, three books" is Roman history, "Adam's Roman Antiquities" is Roman manners and customs, "Graeca Majora" is an anthology of canonical scholarship in Greek, and "Fluxions" is calculus

The "Yale Report" was the Yale faculty's passionate attempt to justify continuing with this curriculum, which in Yale's case wasn't 100% totally wedded to the far-distant past as was sometimes claimed—with the faculty having added, for example, symbolic algebra, logarithms, calculus, and political economy to its roster, which were relatively modern fields of study for the era. But a large portion of it was based on antiquity, and it certainly was a rigorous course of study to most students of the day, with many, if not a majority, growing to eventually despise it.[3] The more recent mathematical courses in the curriculum only compounded the difficulty, and people were quite justified in questioning the relevance, arrangement, difficulty, and manner in which the entire course of study was taught: just as they often are with the curricula of today.

The Yale faculty contended that the essence of the classical curriculum spoke to many recurring features of professional life, and though not always directly applicable to other areas, the broad principles, techniques, and "strenuous exercise of the intellectual powers" honed by this course of study embraced what was "common to them all."[4] Although the classical curriculum of 1828 eventually gave dramatic way to a far more diversified set of course offerings and majors by the century's end, its most idealistic leanings of using a broad approach to education—through illuminating and exploiting the connections between a wide array of disciplines and life in general—is still alive and well in some circles today.

And here is the same desire, embedded in the assertions cited at the beginning of the chapter by Keyser and Thurston, though they are separated by nearly a century. The mathematician Alvin White called the application of this philosophy to mathematics "humanistic mathematics" and helped to jump start a robust network for this way of thinking about math as a human endeavor in the late twentieth century and beyond. The refrain behind this way of thinking is that many important rhythms of life as experienced both individually and collectively by human beings manifest themselves in a wide array of subjects—including mathematics—and that these different manifestations should be deliberately sought out, identified, and connected together. This pedagogical philosophy represents one of the grandest of all the grand confluences in education, acting much like the conceptual superhighway from Chapter 8 to connect big ideas across a wide range of disciplines—both inside and outside of the curriculum.

Though proponents and critics alike believe that an educational curriculum at any level should have a significant impact on the students passing through it, they vehemently disagree on the particulars of what that curriculum should be or how to devise it. Specifically in the case of algebra, some passionately say that the significant and impactful thing that is happening to students nowadays is actually in reverse of what it should be, believing that it harms those who are forced to study a subject that will be largely irrelevant to their professional lives and may prevent them from advancing or completing their education. They remain firm in the stance that it simply must be cast aside as a general requirement, and replaced by something else.

Others defend algebra's current place in secondary and adult education and the methods used to teach it, insisting that it is a necessary standard to develop students' technical capacities, and that the problem with algebra today is a dearth of qualified teachers, not the subject itself or standard instructional methods.

Still a third group agrees with the second that algebra should remain a required subject but is sympathetic to the concerns of the first, arguing for reforms to address the serious philosophical and infrastructural problems in math education, such as a mountain of topics and an overemphasis on manipulating symbols, testing, and irrelevant practice problems. In their view, simply finding better-qualified teachers is not enough to fix a fundamentally flawed situation, and we should focus our energies on developing new and compelling pedagogical methodologies that make the subject far more understandable and appealing to the majority of students required to take it—methodologies that reflect the true potency and beauty in the subject. This is also a hope often expressed by more than a few mathematics professionals over the past two centuries.

To use a crude military analogy, the first group's stance is somewhat akin to abandoning troop positions in ambiguously held territory and redeploying them to more secure locations, in the belief that adequate supply lines do not exist and can never be established. Thus, making the current positions too dangerous for the majority of troops, save, perhaps, for a comparatively few specialized units. Those who argue for improving status quo methods can be viewed as wanting to maintain these positions in the belief that adequate supply lines do exist, but are failing due to incompetent or inadequately trained staff and troops. The third camp, advocating for reform, accordingly believe that by reinforcing the troops and creating additional, alternate routes within a preexisting broad network of supply lines, their position need not be abandoned—and, moreover, can be strengthened. Keyser's and Thurston's statements most naturally align with those in the third camp.

Viewed purely from a military point of view, all three possibilities may make sense depending on the circumstances in which an army in the field finds itself. But what of the educational circumstances that legions of students find themselves facing in an algebra class?

I have adopted the viewpoint of the third camp and have devoted a substantial portion of this book to opening such pathways into existing supply lines for algebraic illumination, insight, and understanding. In this concluding chapter, we'll summarize some of the forms that these paths have taken, but it remains for you to judge for yourself whether these attempts have been successful.

RELATIONSHIPS, SYMBOLS, AND MANEUVERS

Just as arithmetic gives us a precise quantitative vocabulary to express notions such as one collection being larger than another collection, algebra allows us to give shape and personality to our interactions with numerically variable quantities. This is not insignificant, taking full advantage of certain capital features of nature. Consider that though one can write the name of a town with pencil, ink, or chalk, it is also possible to spell the same word using toothpicks, bricks, shrubbery, contrails from an airplane, sand, tape, dominos, people, Play-Doh, and so on. Although its letters can be formed from dramatically different materials—even those not designed for writing—the order of the letters and relationships between them are paramount and are what continue to allow us to recognize the word.

Relationships that we establish between objects, concepts, and quantities can translate to a broad spectrum of materials and circumstances, enabling us to connect them to a broader network of ideas and scenarios. Algebra is inherently relational, helping us to express, maneuver, and transform a wide array of quantitative relationships in extremely efficient and often emergent ways. Two of the most important ways in which algebra does this have been characterized in this book as the two dramas, with the first drama roughly corresponding to what can be captured by variable expressions and the second drama corresponding to what is capturable as equations. A third way—involving graphical representations—has not been critically examined in this book, but is nevertheless extremely important in its own right.

In the first two dramas, relationships between numerically varying quantities are translated into symbolic relationships between the letters that represent them, thus rendering variable situations into a visual and operational format. This represents another grand confluence because once we take variable phenomena and capture them through the lens of visible symbols, it allows us to sometimes make surprising connections in algebraic writing between situations that on the surface may not look similar at all, as occurred in the section "Algebraic Songs" in Chapter 7 and throughout Chapter 8.

These translations to symbolic representations can be likened to the way spectral analysis allows us to identify the constituent material components of astronomical objects by analyzing the light they emit or absorb—or to how detectives can link a person to the scene of a crime by comparing the unique ridges and curves of their fingerprints with those at the scene. In algebra, it is the visible symbols and the relationships they express that facilitate these connections. The transfer of variable situations into their notational renderings really do serve as a type of algebraic fingerprinting.

But algebra offers more than just the symbolic fingerprinting of variable behavior. Consider that, of all the physical materials we mentioned to form a word, the Play-Doh variant is probably the most pliable and easiest to manipulate. Play-Doh has memory and can represent and hold to a certain shape, but it is also malleable and can easily be molded into other shapes. Algebra and arithmetic as we practice them today are similar insofar as the symbols we use can both hold a certain form and also be conveniently reconfigured into other useful forms.

For instance, in arithmetic, we may have six collections of 22, 57, 28, 45, 23, and 25 items, respectively, whose sizes we are able to write down with numerals—and that may be all that is needed, depending on what we want to do with the information. In this case, the symbols act as an aid in remembering the amount in each collection. However, these numerals contain far more content than just the ability to record information. As we know, they can be combined and maneuvered to give new insights. We can start by setting up the addition of $22 + 57 + 28 + 45 + 23 + 25$. This can, of course, be calculated in the traditional vertical fashion, but we can also reorganize the problem as $(22 + 28) + (57 + 23) + (45 + 25)$, which allows us to even more quickly condense the expression into $50 + 80 + 70$ to obtain 200. This computational feature, where the numerals transform—according to certain well-established rules—into new values that match what we actually observe, is an example of when the symbols act in a malleable fashion.

It was a revolutionary discovery in arithmetic that certain types of numerals could actually be made to conveniently and reproducibly transform in this way. That is, these numerals were not static vessels for storing information, but dynamic vehicles that could be maneuvered by users into making nontrivial, extensive calculations singly on their own in writing unassisted by a mechanical instrument. For most ancients, notations such as Roman numerals and hieroglyphic numerals were primarily used to record quantitative information with the calculations being performed on a computing device like an abacus, but the ancient Indians found a way to unify both information storage and calculational functionality into one set of numerals—the Hindu-Arabic numerals—and the rest, as they say, is history.

A similar and equally dramatic revolution occurred in algebra, but it took mathematicians centuries to come to this realization about the possibilities with the symbols used there, with the old rhetorical methods expressed in words giving dramatic way to the more operational symbolic ones developed in the watershed sixteenth and seventeenth centuries.

Firstly, modern symbolic algebra inherits all of what representational and computational arithmetic offers, then scales and intensifies it to represent infinitely many possible versions of an arithmetic statement or action by a single grand variable expression. Secondly, it allows us to capture and separate out all of the different types of variations in a situation, combining those that are the same and separating those that aren't, simultaneously managing variation on multiple channels. When handled properly, these procedures of algebra become strategic and decisive deployments not just pedestrian manipulations, all of which imbue algebra with the power to process and organize the symbolic representation of many types of variations that occur in nature—rendering it indispensable to fields such as science, business, statistics, and data science.

AT THE ALGEBRAIC SPORTS BAR

On Saturday afternoons in the fall, sports bars across the nation may have on as many as a dozen or more different Division I college football games, each one flavored by the history, traditions, and unique fan bases of both teams on the field. Some of these games will be remarkably one-sided affairs, as a powerhouse team takes on a much weaker opponent, whereas others will go down to the wire and may even result in an upset.

Yet, despite the variety of possible matchups and outcomes, all of the games will still have enough in common that we can easily recognize them all for what they are—college football games. Changing channels from game to game, we can see that in spite of the variation between contests, they each feature 60 minutes of regulation play, a system of downs, a line of scrimmage for each play, players clad in protective helmets and pads with numbered jerseys, referees, a brown ball with pointed ends, and other distinctive rules and positions. The simple ability to change TV stations from game to game, or to view them on multiple screens, provides a sports example of the dynamic interplay that is possible between variable elements at work around a strong core of stability.

This is but one illustration of many such productive interactions between variation and stability that occur in many aspects of nature and society—others include the following:

- **Spoken languages:**
  - **Variables:** There are more than 6000 spoken languages in active use worldwide.[5]
  - **Constants:** Languages share the common goal of facilitating structured communication with words that convey meaning in speech, gestures, or writing.
- **People:**
  - **Variables:** The global human population numbers in the billions, but each person is an individual with a unique personality, physical characteristics, talents, skills, and ambitions.
  - **Constants:** Humans are characterized by common anatomical features, including a complex brain, heart, bipedalism, and opposable thumbs, among other mammalian traits.
- **Nations:**
  - **Variables:** There are nearly 200 different nations in the world.

- ○ **Constants:** Every nation has a territory, population, laws, and a system of government.
- **Automobiles:**
  - ○ **Variables:** There are hundreds of different models of automobiles in service today.
  - ○ **Constants:** Automobiles share many common characteristics, including an engine or motor, energy source, brakes, and tires.

Sometimes, this intricate dance between variability and stability can be of a much more quantitative nature, which can help us articulate more detailed and precise mathematical descriptions. In this book, we've discussed examples that include

- businesses, with varying costs and revenues, yet common methods to represent and locate their break-even situations;
- course instructors, with varying assignment categories and weights, yet common methods to represent and compute their course grade averages;
- students, with varying course loads and academic performance, yet common methods for schools to compute their grade point averages;
- investors, with portfolios of diversified investments at varying levels of appreciation or depreciation, yet common ways to calculate their overall return on investment.

Algebra provides a powerful way to comprehensively treat these latter quantitative situations, using symbols to organize and coordinate the variable components and the more stable ones. We've identified these more consistent or stable components in this book as scenario variables—stable within a given scenario, but variable from scenario to scenario.

The existence of scenario variables and the need to distinguish them from traditional unknowns was probably the most valuable of the major advancements in algebra during the 1500s—which is saying something, as it was a banner century for algebra. As we discussed in Chapter 4, scenario variables are more commonly called parameters and were definitively introduced in the late sixteenth century by French mathematician François Viète.

A little over 40 years later in 1637, French mathematician and philosopher René Descartes modified Viète's ideas in his treatise *La Géométrie* by using letters early in the alphabet for parameters and letters later in the alphabet for traditional unknowns, or regular variables.[6] Though Descartes protocol is often followed in elementary algebra instruction, it can be much more inconvenient to adhere to in ensuing applications, as we saw in Chapter 8.

These two types of variable quantities—parameters and regular variables—allow us to transform single-instance algebra into what we have termed "big algebra," illustrating how algebraic insight gleaned from one problem can be scaled to a wider range of applications. For instance, to find the break-even point for various businesses, we may have to solve an individual equation such as $5x = 3x + 20000$ or $752x = 534x + 987000$ or $15.65x = 7.82x + 72000$. As we witnessed firsthand in Chapter 4, parameters then enable us to represent all three of these equations, and thousands more like them, by the single all-encompassing equation $Px = Cx + F$, where P, C, and F take on particular constant values in the context of a specific problem, but change values to reflect different scenarios. This single equation with parameters establishes a holistic, structural connection between all break-even problems that share the same constraints, and individual equations specific to a given scenario become particular instances of that general equation. For example, in the first and second cases, respectively, we have $P = 5$, $C = 3$, $F = 20000$ and $P = 752$, $C = 534$, $F = 987000$.

We considered a similar case in Chapter 8 where we saw that for the three assessment categories of student homework, tests, and final exam scores, we could represent all course average calculations with the single big algebra formula $ax + by + cz$. Here, the letters a, b, and c serve as parameters and capture the contribution values—or weight—of each category, which remain consistent for a particular instructor in a given class, but may change from class to class or from instructor to instructor.

Moreover, algebra is capable of far more than only containing a multitude of situations in a single formula. Like Play-Doh, it can render variable expressions malleable enough to maneuver and transform them into radically different shapes and forms that generate novel insights. One especially pronounced instance of this capacity is visible in quadratic equations, where we can represent a galaxy of such equations with the single big algebra formula $ax^2 + bx + c = 0$. From this we can apply the technique for solving each such individual equation, once and for all, to this single super-equation and derive the famous quadratic formula:

$$x = \frac{-b \pm \sqrt{b^2 - 4ac}}{2a}.$$

(See Appendix 1 for more details.) This formula holds the collective results of infinitely many individual acts in suspended animation, showing through the spectacular use of the parameters—a, b, and c—the final form that all of the specific instances can end up taking. We also saw the potential of big algebra in the various interpretations of the bill denominations problem from Chapter 11.

Significant scientific laws and processes have historically been captured by algebraic formulas capable of handling a wide array of different inputs—including mass, charge, the coefficient of friction, the spring constant, electrical resistivity, and electrical conductivity—through the use of parameters. Similarly, in real-life financial situations, parameters can stand in for fixed numerical information such as the sales tax in various towns, the interest rates at a given time, a principal amount invested, the price of gas per gallon on a given day, or the hourly wage of various workers. Although these constant quantities may be fixed in a particular location, for a certain period of time, or concerning a specific person, they can vary depending on the unique circumstances of the problem. In tuning our parameters from scenario to scenario, we become like patrons at the algebraic sports bar where—instead of changing channels between football games—we change channels from scenario to scenario identifying their similarities and differences, their constants and variables.

A HIGHER CALLING

As breathtaking as it may be at first sight, the Grand Canyon becomes all the more impressive once we understand its dimensions, what it contains, and how it was slowly carved out by the Colorado River over millions of years. The realization that some of the same smaller-scale processes at work in the erosion you witness in your backyard or at a local municipal park along the river also created this immense natural wonder is a spectacular and powerful grand confluence of ideas.

Emergent viewpoints in geology—such as understanding the relationship between micro and macro processes—makes much of what we observe on Earth more comprehensible and less mysterious. Moreover, these viewpoints provide even nonexperts, who have some understanding and awareness of them, an entire framework within which to place and analyze new and unfamiliar landscapes that they happen upon. Such awareness can also lead individuals to acquire a greater appreciation and awareness of their natural environment, which may ultimately lead them to become even less intimidated by the subject of geology itself. The same can be said for astronomical knowledge. As Bernard de Fontenelle stated, "Nature…is never so admirable, nor so admired as when it is known."[7]

Consider how we might apply the same principle to large swaths of basic numeration and even some aspects of geometry. Most people can tell you very little about the nuances of land surveying but, at the same time, are often quite comfortable with basic geometric concepts such as length, angles, area, and volume. Moreover, those same people can successfully apply those concepts to calculating the distance between two towns, the heights of people or structures, the square footage of a house or room, or the area of a garden.

How might we frame such a relationship between micro and macro algebra? What would it look like—and is it even possible—for someone with a rudimentary understanding of variable quantities, and a passing familiarity with algebra from a distant course or two, to extrapolate that basic proficiency to new situations involving variable phenomena? Variation and all that it entails is a much more abstract idea, evidenced by the fact that most people readily internalize elementary arithmetic, whereas conceptual retention of fundamental algebraic principles is orders of magnitude lower in the general population.

Algebra in a sense does for arithmetic what arithmetic does for the general notions of the size of a collection, measurement, and ordering. That is, algebra offers a set of rules for manipulating mathematical symbols that represent objects and operations from arithmetic, a subject that itself offers a set of rules for manipulating symbols that represent numerical quantities. A second level of symbolization and a second level of generalization can be tough to master, but symbols at both levels serve a powerful purpose.

One such power of symbolism is that symbols enable us to speak about things in their absence.[8] This is equally true of drawings, photographs, and maps. In many cases such as with military units, sports, or travel, such representations allow us to better understand the things being represented while probing them for weakness or limitations and gleaning new insights. So too is it with the alphabetic, diagrammatic, and graphical symbols we use in algebra.

As art education scholar Elliot Eisner wrote in *The Arts and the Creation of the Mind*, "Ideas and images are very difficult to hold onto unless they are inscribed in a material that gives them at least a kind of semipermanence."[9] Yet, how can we present algebra in such a way that its symbols and methods illuminate and expand our mathematical knowledge, rather than obscure and obstruct it? This is one of the central problems in algebra education—and it has proven to be a tougher nut to crack than for arithmetic.

What do people remember about or take away from the other classes they take? Though undoubtedly not as much as their teachers might hope, students do perhaps take general lessons away with them into their future coursework and professional lives. They may even learn some of these lessons outside of class.

Take science classes, for instance. Former science students are likely to be at least somewhat familiar with the concepts of atoms, planets, stars, power, and forces, as well as physical phenomena they personally can experience like electricity, gravity, velocity, and acceleration. Some can also recall the names of a few of the famous scientists associated with major discoveries. From chemistry, even a student who has forgotten how to balance an equation is likely to know that mixing unknown substances together could produce dangerous gases or even cause explosive reactions. From biology, they know that tiny microorganisms can carry contagious diseases and that airborne pathogenic microbes can infect us.

What do people think about when they recall algebra? Common responses may involve letters, equations, and manipulations, but rarely with an appreciation of what those elements are really about. If students leaving algebra classrooms can't articulate why algebra is conceptually significant, even at the most basic level, this indicates a breakdown in algebra education relative to other subjects in the secondary school or college curriculum—something that has made it a topic of ongoing discourse in math education. This book has been an attempt to contribute a few ideas to that discourse while simultaneously fostering a greater appreciation of the magic and wonder of algebra in general— one of many possible higher callings of algebra educators.

William Thurston once wrote that "it is easy to forget that mathematics is primarily a tool for human thought,"[10] and it's true that the perspective that's often lost in translation in the algebra classroom is that algebra can shape the way that we *think*, both inside and outside the context of a specific problem. An algebraic way of thinking seeks a more global and comprehensive understanding of quantitative variable behavior in its various guises, and seeks to consolidate that understanding in productive, formulaic, and operational ways—including variable expressions, equations, tables, and graphs. This way of thinking actually transcends algebra, but the potential exposure to it for most students receives a golden opportunity in an algebraic setting.

Though teaching algebra as a way of thinking rather than as a means to an end may be difficult to achieve in the current educational environment, a student outcome that achieves a more holistic understanding of algebra—and emphasizes the purposes and meaning behind the procedures being taught—would be a step in the right direction and is a worthwhile goal in and of itself. I believe that Daniel Willingham is correct in his claim that, though many of the complaints against algebra involve questions of relevance outside of the classroom, the far greater concern is that many students simply don't understand the rationale behind the calculations they are performing.[11] The unfortunate reality that most of them find no cohesion—no conceptual organizing principle either within the subject itself or in their efforts—doesn't help matters either.

If students understood the purpose of algebra from the beginning, they may leave their studies with both a better feeling toward the subject and a better sense of how they might make use of what they've learned. This is by no means a trivial thing, amounting to what John Dewey calls having a worthwhile experience (as discussed in the introduction).[12] However, this is not a comprehensive picture of what may be possible. Students certainly will encounter quantitative variation elsewhere in life—what would be some of the things that they could think about when they encounter it?

One of the things we've touched on several times throughout the book is that numerical variation in everyday life is not always transparent, and that it can take some work to even be aware of its presence. When we encounter quantitative variation in the wild, it is often as a specific instance of a numerical value. Sometimes this particular value may be all that we are interested in, just like sometimes we only want to know the specific temperature and weather conditions on a certain day.

However, more is available if we want. Just as a person can learn much more about a geographical region by being aware of the range of temperatures and weather over a period of time—its climate—so too can the algebraic way of thinking help us to better understand the climate of a particular phenomenon. This can be done in part by having a better feel for the range of values it can take on; so, when we encounter a particular number for a particular type of behavior, it may be productive to look at that value as a particular instance—like a particular temperature—of something more general. Questions we could ask include: What type of algebraic climate could produce these values? What patterns can we identify that will tell us more about it?

Think of it this way. An alien coming to Earth might first see children, young adults, and the elderly as distinctly different species of people, not understanding the aging process or that one group morphs into another group over time. However, after much observation over time, they could eventually reach this conclusion on their own by studying the relationships and interactions between each group.

Encountering specific instances of variation is a bit like this. Modeling scenarios algebraically gives us the ability to tie together situations that we may not initially understand or see as connected. Alternatively, if a range of values are what is first encountered, then we can go the other way and ask if there is some core mechanism that ties them together in a formulaic way: Algebra's investigative properties work both ways. As we discussed in Chapter 9, part of understanding that core mechanism may include trying to find out if it easily splits into scenario variables and regular variables and identifying what those are. This can be a useful organizing principle even when a detailed, technical understanding is not attainable.

As a mathematician educator, I've found that developing this kind of higher-order algebraic awareness tends to be easier for adult students. Adults are trying to weave much more complicated tapestries in their understanding than children who are still developing; this instinct can be leveraged and built upon through a humanistic approach to mathematics that connects with their own experiences. In my teaching, I aim to give my students a sense of the bigger picture of algebra's contribution to our collective understanding, which can mean giving older students a productive new perspective that alters or evolves the way they see the world and what they already know about it. One of my goals in writing this book has been to capture the magic and impact that such a global shift in perspective has had for my adult students—and though there are many different ways to teach algebra and my approach may not necessarily be the most effective for the high-school classroom, my hope in this final movement of our algebraic symphony is that you will close this book with a better sense of what algebra has to offer us. I may not have convinced you that the algebraic way is "fun," but I hope you'll agree that algebra can be interesting, accessible, and full of possibilities.

Figuring out how to successfully incorporate topics into an algebraic setting can be likened to figuring out how to incorporate various ingredients into cooking a good meal. We know what needs to be added—some ingredients essential and others optional—but it is the proper mixing of these—along with proper amounts of heat and time—that turn our efforts into a good meal or not. As just one cook in a kitchen that features a wide assortment of recipes and cooking styles, I hope you have enjoyed my offerings.

ALGEBRA THE BEAUTIFUL

We've thought about algebra throughout this book as a vast continent of knowledge, containing both areas that we already know fairly well and tracts of less penetrated terrain brimming with unused potential for readers to explore. It is a grand subject whose ability to exploit a remarkable property of nature is key—that property being that it is possible to represent, describe, mimic, and transform a wide array of phenomena using symbolic systems that obey certain protocols.

Paradoxically, it is exactly these protocols that are often looked upon as being among the most monotonous of the features of algebra, no doubt contributing in part to the subject's large PR problem. Yet, it is exactly these procedures—along with the fact that they were acquired to deal with one situation but are capable of being continually reused to deal with other situations—that account for some of the most intense aspects of the beauty of the subject. Imagine a retail store gift card with a zero balance. Such a card can be looked upon as being a valueless rectangular piece of plastic or as an object having a lot of potential. This potential can be realized through recharging or reloading the card by adding money to its balance. Once this has been done, a wide assortment of possible purchases emerges.

Similarly, the symbols and protocols of algebra can be looked upon as being completely devoid of value, the many manipulations just so much arbitrary, meaningless ritual. However, like the empty gift card, these protocols and rituals have the potential to be given immense value. Algebraic expressions and operational procedures can be continually "charged up" to represent and say important things about a wide array of quantitative variations, including the ability to take these charged-up representations and make spectacular transformations of form to find out unknown, new information and insights almost like no other discipline that preceded it or that isn't presently underwritten by it. In other words, the apparently mindless ritual can be electrified to great purpose—acquiring immense worth. In this, it shares great similarities to computer science, natural science, and engineering.

Thus, elementary algebra when viewed from a certain perspective can contain in miniature some demonstration of how these other disciplines contribute to the advancement of human knowledge and our understanding of the universe around and beyond us. It is a subject with a fraught pedagogical history, at one time having its secrets completely masked from public view, like a half-mythical impenetrable forest, and yet at the present time having many of its secrets hidden from the public in plain sight, with most people unable to appreciate the algebraic forest now for its many trees.

Algebra is a heritage that belongs to every one of us—truly one of the intellectual wonders of the ancient and modern world, and though probably not perched as high, by discipline, as one of the top seven such wonders, it certainly has helped bankroll some of those in the top spots.

Regardless of where it sits, algebra is a vast, scenic, wide-ranging continent of possibilities that provides conceptual fuel for mathematics, as well as much of science, engineering, and a host of other disciplines vital to modern life. It was already a very big deal during its modern symbolic ascent back in the age of printing presses, Mercator projections, armadas, Kepler, and slide rules, and it still remains a big deal in this current era of the internet, global positioning systems, carrier battle groups, data science, and computers.

Algebra the Beautiful—fertile, electrical soil indeed.

Appendix 1

The Quadratic Formula and Parameters

The well-known quadratic formula gives an example of the all-encompassing power of parameters. Examples of quadratic equations include the following:

1. $6x^2 + 13x + 8 = 0$.
2. $x^2 + 11x + 5 = 0$.
3. $20x^2 + 5x - 36 = 0$.
4. $\frac{2}{3}x^2 - 7x + 72 = 0$.

Techniques for solving some problems of these types existed going all the way back to ancient Babylon over 3500 years ago—although it was only in the sixteenth and seventeenth centuries that the presentation of these equations would be similar in form to the way that the four examples are written.

One of the general methods for solving these equations is known as "completing the square." The method can be applied to each of these equations to find individual solution(s) for each.

In the language of Chapter 4, we can think of each quadratic equation as describing an individual scenario where the three fundamental types of behavior (x^2 behavior, x behavior, and constant behavior) each have a specific numerical value (or "price in dollars") assigned to them. The values of these "prices" can change from scenario to scenario but remain constant in a given scenario and are what we want to capture as parameters. Note that these "price values" are often called "numerical coefficients" or "given values" or "givens" for a specific scenario.

So, in Equation 1 from the list, the price value for x^2 is 6, for x is 13, and for the constant is 8. Once these price values are fixed, the x is still allowed to take on different values (like it was in the break-even situation from Chapter 4, where it represented the number of meals in a specific scenario).

For Equation 2, the price this time for x^2 is 1, for x is 11, and for the constant is 5. And we get a new scenario, where the x is still free to take on different values. Similar situations occur for Equations 3 and 4.

Just as we were able to use P, C, and F to represent the dollar values of selling price per item, cost to make each item, and fixed costs, respectively, in the break-even scenario, we can use letters here to represent the "price values" for the three terms x^2, x, and the constant. The standard letters to use here are those early in the alphabet: a, b, and c, respectively.

Doing so, we obtain the following equation: $ax^2 + bx + c = 0$. The three letters a, b, and c represent parameters, and what they give us is the power to represent all quadratic equations by a single super-equation. All of the individual equations can be obtained from this one by appropriate choices of a, b, and c. We illustrate this in the following table:

Set Parameters			Super Equation $ax^2 + bx + c = 0$ Becomes:
a	b	c	
6	13	8	$6x^2 + 13x + 8 = 0$
1	11	5	$x^2 + 11x + 5 = 0$
20	5	-36	$20x^2 + 5x - 36 = 0$
$\dfrac{2}{3}$	-7	72	$\dfrac{2}{3}x^2 - 7x + 72 = 0$

Reasoning similarly, we can obtain all of the infinitely many other quadratic equations by appropriate choices of a, b, and c from the general equation.

We are not done, however.

The coup de grâce is that the method of completing the square, which can be used to solve each individual equation in its specific scenario, can now be applied to the super-equation ($ax^2 + bx + c = 0$) to yield a general solution to the infinitely many equations all at once—in one grand maneuver. Doing so in this case ultimately yields the famous quadratic formula (details not shown):

$$x = \frac{-b \pm \sqrt{b^2 - 4ac}}{2a}.$$

The formula represents in writing a crystallization of the entire process of completing the square. It also means that once we identify the parameters in a given quadratic equation (which can be done on sight), instead of having to perform the more involved method of completing the square each time, we can simply plug the values for a, b, and c into the crystallized quadratic formula and do some arithmetic, and the solution pops out for us.

Moreover, this technique of using parameters to freeze-dry in writing the results of many algebraic maneuvers is not limited to the quadratic equation here or the break-even equation in Chapter 4, but can be used in all kinds of other situations with similar effect. This is what we have called big algebra.

Fertilizing the soil for others to purposefully use and systematically apply parameters for wide-scale impact and insight is—in the mind of many mathematical historians—Viète's most important and revolutionary contribution to mathematics.

Appendix 2

Five Word Problems

1. An unknown number added to ten more than fifteen times itself gives one hundred six. Find the number.

   The relationship is (the unknown number) + 10 more than 15 times (the unknown number) gives 106. Tag the unknown number by x. Translated to algebraic symbols, the problem becomes $x + 10 + 15x = 106$. This equation simplifies to $\mathbf{16x + 10 = 106}$.

2. A 106-foot length of rope is cut into three pieces. The second piece is 10 feet longer than seven times the length of the first, and the third piece is eight times the length of the first piece; find the lengths of all three pieces.

   The three pieces are related to each other in the following way:

   - The second piece is 10 feet more than seven times the length of the first piece.
   - The third piece is eight times as long as the first piece.

   Let the length of the first piece be represented by x. If we do this, we obtain the following algebraic relationships:

   - First piece length = x.
   - Second piece length = $10 + 7x$.
   - Third piece length = $8x$.

   We know that first piece length + second piece length + third piece length = 106. Translating to algebra gives the equation $x + 10 + 7x + 8x = 106$. Simplifying the left-hand side gives $\mathbf{16x + 10 = 106}$.

3. Given a rectangle of perimeter twice 53 meters and whose length is five more than seven times its width, find its length and width:

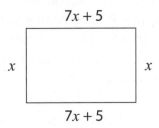

$$7x + 5$$

$$x \qquad\qquad x$$

$$7x + 5$$

The perimeter is given by 2(length) + 2(width). We have the following from the diagram:

- Length = $7x + 5$.
- Width = x.

In the language of x, the perimeter becomes $2(7x + 5) + 2x = 14x + 10 + 2x = 16x + 10$.

The perimeter's quantitative value in this problem measures twice 53 meters, which is 2(53 meters) = 106 meters. This gives the following relationship: Perimeter in x language = perimeter's quantitative value. Expressing this relationship in algebraic equation language yields **$16x + 10 = 106$**.

4. If a power tool rents for $16 a day plus a one-time $10 processing fee, how many days can you rent the tool if you have four $20 bills, two tens, a five, and a dollar bill to spend?

We want to find the number of days that we can rent using the available money. Let x = number of days. The money we spend for the rental is 16 times (the number of days) + $10 processing fee, or equivalently in x language, $16x + 10$.

The available amount of money is 4(20) + 2(10) + 5 + 1 = 80 + 20 + 5 + 1 = 106. This gives the following relationship: Money spent in x language = available amount of money. Expressing this in algebraic language yields the equation **$16x + 10 = 106$**.

5. Next Saturday Barbara will be doing a job for a client that pays her $36 an hour. The job requires specialized computer services that cost $20 an hour to use in addition to a $40 setup fee. When she arrives on location for the job, she notices a $50 bill that the client left to say thanks for coming in on the weekend. How many hours does she need to work so that her total profit (including her tip) for the day is $106?

We want to find the number of hours Barbara needs to work to yield a profit of $106. Let x = number of hours. Profit is calculated as Barbara's earnings minus her costs. Here are those values in x language:

- Barbara's earnings = (hourly wage)x + gift = $36x + 50$.
- Barbara's costs = (hourly charge)x + setup fee = $20x + 40$.

Barbara's quantitative profit in dollars equals $106.

This gives the following relationship: Barbara's profit in x language = Barbara's quantitative profit. Expressing this relationship in algebraic symbols yields

$$\underbrace{(36x+50)}_{\substack{\text{Barbara's} \\ \text{earnings}}} - \underbrace{(20x+40)}_{\substack{\text{Barbara's} \\ \text{costs}}} = 106.$$

Simplification gives $36x + 50 - 20x - 40 = 106$ or $\mathbf{16x + 10 = 106}$.

Appendix 3

Function Notation [y and f(x)]

For those familiar with function notation in school algebra, in the situation when both sides of an equation come into play, deciding whether to use a single-letter abbreviation or not is often faced: that is, deciding whether to use $f(x)$ notation versus the letter y. Once again, the viewpoint is more one of user-friendliness than of necessity.

For instance, sometimes you may see the following two representations: $f(x) = 16x + 10$ or $y = 16x + 10$. These both represent the same variation given by $16x + 10$, but they indicate two different ways of looking at the situation. The first option looks at the function $f(x)$ as being dependent in a subordinate fashion on the terms on the right-hand side. That is, we can plug in $x = 2$ to get $f(2) = 16(2) + 10$, which simplifies to $32 + 10 = 42$.

However, if we want to discover what values of x will make $16x + 10$ equal 106, 586, or 938, we tend to look at the second representation, $y = 16x + 10$. In this case, both sides of the equation are in play with equal authority. Thus, we would want to find what x values correspond to $y = 106$, 586, and 938, respectively. The easiest way to do this is to first solve this equation in one grand maneuver for x. The reduction diagram shows how this unfolds:

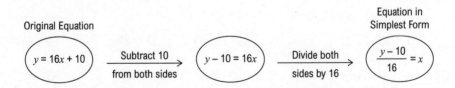

Original Equation

$y = 16x + 10$ — Subtract 10 from both sides → $y - 10 = 16x$ — Divide both sides by 16 → $\dfrac{y - 10}{16} = x$

Equation in Simplest Form

Now, if we place the given y values in the simplified equation, we get the corresponding x values of 6, 36, and 58. Here is the calculation of x when $y = 938$:

$$x = \frac{y - 10}{16} = \frac{938 - 10}{16} = \frac{928}{16} = 58.$$

Similarly, we will obtain $x = 6$ for $y = 106$ and $x = 36$ for $y = 586$.

If we had used $f(x)$ here instead of y and solved for x, in this notation we would have obtained

$$x = \frac{f(x) - 10}{16}.$$

This form, though accurate, is generally less user-friendly to new learners of algebra, and probably so even to most of those well versed in algebra.

For those familiar with graphing, when we look at graphing ordered pairs of x and y on the Cartesian coordinate system, we often view the representation $y = 16x + 10$ as being more convenient because we have the two axes (x-axis and y-axis) in play.

Again, these are subtle viewpoints; though they are often employed in common practice, they are not absolutely necessary.

Appendix 4

Exponential Functions of Different Bases

To see how exponents of one base can yield exponents of other bases, we need to make use of an important property of exponents that is often called the *power rule*. To see how it works, let's look at a few examples of exponents raised to an additional power.

- $(2^3)^2$: This means to raise 2^3 to the second power, so we have a power raised to another power.

  Doing so yields $(2^3)^2 = (2^3)(2^3)$.

  This equals $(2 \times 2 \times 2)(2 \times 2 \times 2)$ because 2^3 means to multiply three 2s together.

  Thus, $(2^3)^2 = (2 \times 2 \times 2)(2 \times 2 \times 2)$.

  Dropping parentheses gives

$$\left(2^3\right)^2 = \underbrace{2 \times 2 \times 2 \times 2 \times 2 \times 2}_{\text{six 2s multiplied}}.$$

  This means that $(2^3)^2$ is the equivalent of multiplying six 2s together.

  Thus, $(2^3)^2$ equals 2^6 or $(2^3)^2 = 2^{3 \times 2}$. (This is the power rule.)

- $(5^4)^3$: This means to raise 5^4 to the third power.

  Doing so yields $(5^4)^3 = (5^4)(5^4)(5^4)$.

  This equals $(5 \times 5 \times 5 \times 5)(5 \times 5 \times 5 \times 5)(5 \times 5 \times 5 \times 5)$ because 5^4 means to multiply four 5s together.

  Thus, $(5^4)^3 = (5 \times 5 \times 5 \times 5)(5 \times 5 \times 5 \times 5)(5 \times 5 \times 5 \times 5)$.

  Dropping parentheses gives

$$\left(5^4\right)^3 = \underbrace{5\times5\times5\times5\times5\times5\times5\times5\times5\times5\times5\times5}_{\text{twelve 5s multiplied}}.$$

This means that $(5^4)^3$ is the equivalent of multiplying 12 5s together.

Thus, $(5^4)^3$ equals 5^{12} or $(5^4)^3 = 5^{4\times3}$. (This is the power rule.)

- **$(2^2)^x$:** Using the power rule, we have $(2^2)^x = 2^{2 \cdot x} = 2^{2x}$.

  But we also know that $(2^2)^x = (4)^x = 4^x$ (because 2^2 equals 4).

  Thus, $2^{2x} = 4^x$.

  Note that 2^{2x} has a base of 2 and exponent of $2x$.

  This means that 4^x can be written as an exponent that has a base of 2.

- **$(2^3)^x$:** Using the power rule, we have $(2^3)^x = 2^{3 \cdot x} = 2^{3x}$.

  But we also know that $(2^3)^x = (8)^x = 8^x$ (because 2^3 equals 8).

  Thus, $2^{3x} = 8^x$.

  Note that 2^{3x} has a base of 2 and exponent of $3x$.

  This means that 8^x can be written as an exponent that has a base of 2.

- **$(2^5)^x$:** Using the power rule, we have $(2^5)^x = 2^{5 \cdot x} = 2^{5x}$.

  But we also know that $(2^5)^x = (32)^x = 32^x$ (because 2^5 equals $2 \times 2 \times 2 \times 2 \times 2$ or 32).

  Thus, $2^{5x} = 32^x$.

  Note that 2^{5x} has a base of 2 and exponent of $5x$.

  This means that 32^x can be written as an exponent that has a base of 2.

Here, we have dealt with numbers that have nice relationships with respect to the base of 2 in that they are powers of 2 (second, third, and fifth powers, respectively). But it is also possible to raise numbers to decimal powers that are irrational numbers, too.

The detailed explanation of this is beyond the scope of this book. We offer the following without proof:

- **$(2^{3.3219280948\ldots})^x$:** Using the power rule, we have

$$\left(2^{3.3219280948\ldots}\right)^x = 2^{(3.3219280948\ldots)\text{ times } x} = 2^{(3.3219280948\ldots)x}.$$

But we also have that $\left(2^{3.3219280948\ldots}\right)^x = (10)^x = 10^x$ (because $2^{3.3219280949\ldots}$ equals 10, the demonstration of which is beyond the scope of this book).

Note that

$$\underbrace{2^3 = 8}_{3} \text{ is less than } \underbrace{2^{3.3219280948\ldots} = 10}_{3.321980948\ldots} \text{ is less than } \underbrace{2^4 = 16}_{4}.$$

$$3 \quad < \quad 3.321980948\ldots \quad < \quad 4$$

Note that $2^{(3.3219280948\ldots)x}$ has a base of 2 and exponent of $(3.3219280948\ldots)x$.

This means that 10^x can also be written as an exponent that has a base of 2.

Observe that the number given by $3.3219280948\ldots$ is a non-repeating decimal that goes on forever. We can drop the "goes on forever" and get as close an accuracy as we want by taking a sufficient number of digits. Note that if we round this off to nine digits, we have 3.321928095, and $2^{3.321928095} \approx 10.0000000008$.

We can do this same process for an exponent to any base and thus any exponential function to any other base can be written in the form that uses 2^{ax}. Here, a is a scenario variable or parameter. We list the various settings of a for some scenarios here:

Scenario	Value of Parameter a	2^{ax}
4^x	$a = 2$	2^{2x}
8^x	$a = 3$	2^{3x}
10^x	$a = 3.3219280948\ldots$	$2^{(3.3219280948\ldots)x}$
16^x	$a = 4$	2^{4x}
32^x	$a = 5$	2^{5x}

The same reach holds true for an exponent whose base is the number e. Using the same letter for the parameter, we will have e^{ax}. Here, the number e takes the place of 2 in the previous examples. The following table lists (without proof) the various settings of a in this situation for the scenarios from the previous table:

Scenario	Value of Parameter a	e^{ax}	a Rounded to Nine Decimals
4^x	$a = 1.386294361\ldots$	$e^{(1.386294361\ldots)x}$	$a = 1.386294361$
8^x	$a = 2.079441541\ldots$	$e^{(2.079441541\ldots)x}$	$a = 2.079441542$
10^x	$a = 2.302585092\ldots$	$e^{(2.302585092\ldots)x}$	$a = 2.302585093$
16^x	$a = 2.772588722\ldots$	$e^{(2.772588722\ldots)x}$	$a = 2.772588722$
32^x	$a = 3.465735902\ldots$	$e^{(3.465735902\ldots)x}$	$a = 3.465735903$

Note that all of the a values listed here are irrational numbers, meaning they have infinite non-repeating decimal expansions and would need to be rounded off for use.

Appendix 5

Solving for *y* in the Equation
12*x* + 8*y* = 400

The following reduction diagram shows how to solve the equation $12x + 8y = 400$ for y:

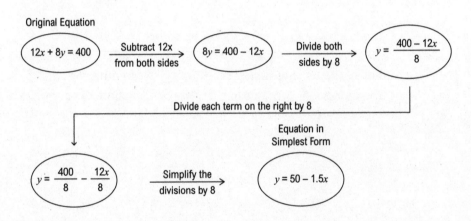

Glossary of Terms

Entries marked with an asterisk are not commonly used terms in the general literature and correspond to concepts whose definitions are more specific to their use in this book.

big algebra*: Algebraic operations applied to expressions or equations that contain parameters, meaning that they are really multiple algebraic expressions or equations represented by a single general one.

break-even point: The point in a business where the revenue brought in equals the overall amount spent.

classroom word problems*: Manufactured problems not likely to be encountered outside of a mathematics education or recreational math context, or possible real-world problems with intentionally sculpted scenarios that may not be as realistic as they would be in actual practice.

cloud of possible values*: All of the possible numerical values that a variable expression can take on from its allowable input values; also known as the range of an expression. A synonym is "symphony of possible values."

conceptual fuel*: Concepts, experiences, and ideas that can be put to effective use in a directly related area or in unrelated areas; something that provides conceptual enlightenment or sustenance.

conditional equation: An equation whose truth depends on the values chosen.

counting numbers: The numbers 1, 2, 3, 4, 5,…; in other words, the whole numbers without 0 or the integers greater than 0. They are also known as the natural numbers.

Descartes protocol: The algebraic protocol that represents regular variables or unknowns by letters late in the alphabet (such as x, y, z) and parameters by letters early in the alphabet (such as a, b, c). This protocol is still extant—though inconsistently applied—and was discussed by René Descartes in his 1637 book, *La Géométrie* (*The Geometry*).

Diophantine equation: Any equation, usually in several unknowns, that is studied in a problem whose solutions are required to be integers, or sometimes more general rational numbers (Nelson 2008).

first drama*: (a) Situations related to a numerical variation or numerical symphony or its capture as an algebraic expression, and the ensemble of all possible values generated by such a variation or expression. (b) Situations involving a function and its domain (all values that can be input) and range (all possible output values), or the relationship between the two.

identity equation: An equation that is always true in its domain of applicability regardless of the values chosen; often called an identity.

indeterminate equation: An equation for which the unknown or variable quantities can have many different values—sometimes infinite in number—that satisfy or solve the equation.

integers: The positive and negative whole numbers:…, -4, -3, -2, -1, 0, 1, 2, 3, 4,…

isotope: One of several varieties of the same element. Elements are identified by the number of protons in the atomic nucleus: If two atoms have a different number of protons, then they represent different elements. The number of neutrons in the nucleus of an element, however, can vary, which leads to different varieties of the same element. For example, the most commonly occurring type of nitrogen atom has seven protons and seven neutrons (nitrogen-14), yet another type of nitrogen atom has seven protons and eight neutrons (nitrogen-15). When two types of atoms of an element have different numbers of neutrons (such as nitrogen-14 and nitrogen-15), they are isotopes of that element. Some isotopes of an element may be stable while others can be radioactive.

multiplication of variables and values:

cross: The symbol $\times$ used to indicate multiplication: for example, $2 \times 5 = 10$.

dot: The symbol $\cdot$ also used to indicate multiplication. We may represent "2 times 3" or "8 times x" respectively as $2 \cdot 3$ and $8 \cdot x$. The dot symbol is often used when the traditional cross symbol $\times$ could be confusing due to its similarity to the letter x, which is often used in algebra.

juxtaposition: Algebra affords us the convenience of using juxtaposition as multiplication. For instance, $x \cdot y$ can be represented more succinctly as xy. This is not advisable using numerals alone because replacing 2×3 by the juxtaposition 23 gives two entirely different values (6 in the former and 23 in the latter).

asterisk: The symbol * is often used to represent multiplication in computer programming languages and software, for example, Excel or some computer algebra systems.

number of days and age problem*: Problem introduced in Chapter 1 as "Magical Three-Digit Numbers" and analyzed in detail in Chapter 2:

> Pick the number of days you like to eat out in a week (choose from 1, 2, 3, 4, 5, 6, 7). Multiply this number by 4. Then add 17. Multiply that result by 25. Next add the number of calendar years it is past 2013. Now if you haven't had a birthday this year, then add 1587, but if you have had a birthday this year, then add 1588. Finally, subtract the year that you were born from this. Reading the resulting three-digit number from left to right, the first digit is the number of times you like to eat out in a week and the last two digits are your age.

numerical symphony*: *See* symphony (numerical).

operative symbolism: A symbolism that is more than just a convenient short-hand. It does more than store information, but can allow manipulation and transformation according to well-defined rules to reproducibly yield new and different symbolic forms that remain useful—in that they may reveal new information and interpretations. For example, the addition of three hundred fifty-two plus one hundred forty-six is performed by first rewriting it in numerals as 352 + 146. Once in numeral form, systematic rules can be applied to yield the new value represented in symbols as 498. Rules such as $2 + 6 = 8$, $5 + 4 = 9$, and $3 + 1 = 4$ still hold when we add together other numerals such as 523 + 461, allowing for the systematic computation of yet another new value represented in symbols as 984. The numerals used in this way form an operative symbolism.

parameter: An object that has aspects of both a constant and a variable. It acts as a constant within a given scenario but can change value from one scenario to another scenario, then taking on the characteristics of variable. It is often referred to in this text as a scenario variable and more generally is sometimes referred to as a coefficient, given, or known.

quantitative cocktail*: A numerical measure whose value depends on a systematic mixture in varying strengths of other, more basic measures; a numerical measure that can be broken up into a discrete set (spectrum) of more basic measures of varying contribution strengths (or intensities).

real-world problems/applications: Quantitative problems that might naturally occur outside of a mathematics class or a recreational math book.

rhetorical algebra: A type of algebra where the primary algorithms are expressed in a non-operative symbolism, usually in the words or word abbreviations of a language such as Arabic, Chinese, Sanskrit, Latin, Italian, German, or English.

rhyme (common): Two or more words or phrases that end in the same sounds.

rhyme (metaphorical)*: Two or more different situations that share a framework of similarity, regularity, or repetition. For example, (a) the expressions 16h, $16h^2$, $16h^3$, and $16h^4$ can be viewed as metaphorically rhyming in the sense that they are different but represent the similar idea of the variable h being raised to various powers; and (b) the witty saying "History never repeats itself, but it rhymes" implies that situations in history are never exactly the same but share important and deep similarities. This is similar to Alfred North Whitehead's definition of *rhythm* as the conveyance of difference within a framework of repetition (*The Aims of Education*, p. 17).

second drama*: (a) Situations related to finding specific input value(s) of a numerical variation (or its capture as an algebraic expression) that make the variation (expression) take on particular output value(s). (b) In algebra, situations involving finding a particular (domain) value of a function that makes the function take on a specific (range) value. The algebraic treatment of such situations usually involves solving equation(s).

symbolic algebra: A type of algebra where the primary algorithms are expressed in an operative symbolism (usually involving letters from a language alphabet such as Latin or Greek but can involve words or abbreviations in any written language).

symbolic maneuver*: Any introduction, combination, movement, and/or manipulation of symbols (including tables and diagrams) to gain an advantage in knowledge, insight, organization, identification, clarity, efficiency, etc.

symphony (music): A lengthy form of musical composition for orchestra, normally consisting of several large sections or movements often tied together around a central theme or emotion.

symphony (numerical)*: An ensemble of varying numbers, expressions, or objects that are tied together around a well-defined or easy-to-recognize procedure, rule, or theme.

term: A product of numbers and variables: for example, $6x$, $6x^3y^5$, and $-90x^4yz^7$. The expression $8xy + 40w^{10}y^5$ contains two terms ($8xy$ and $40w^{10}y^5$). In some cases, such as in the three-term expression $3x + 4x + 25x$, the terms can be combined to become a simpler expression (the single-term $32x$ in this case).

unknown: A yet-to-be-determined value (or set of values) that satisfy a condition (or set of conditions). The algebraic representation of such conditions is usually in the form of equations or inequalities. Variables in an algebraic expression are often called unknowns whenever the algebraic expression becomes part of an equation. It is often used interchangeably with "variable." For example, in the algebraic expression $5x + 6$, the x is viewed as a variable that can take on infinitely many values. However, when placed into the equation $5x + 6 = 66$, it is not uncommon for the x to be viewed as an unknown whose value is to be determined (which is 12 in this case).

variable: A quantity that can take on different values. In algebra, such quantities are often incorporated as part of a larger expression. Variables are useful in modeling changing quantities in nature or human situations. *See also* unknown.

variable (regular): A quantity (when viewed together with parameter quantities in an algebraic expression) that can take on different values within a given scenario (while the parameter is fixed). *See also* parameter.

variable (scenario): *See* parameter.

variable expression: An algebraic expression that contains at least one variable quantity, meaning that the expression itself can vary in value.

Viète/Harriot protocol: The protocol that represents regular variables or unknowns by vowels (such as A, E, or I) and parameters by consonants (such as B, G, or D). Francois Viète introduced this protocol using capital letters in his 1591 book, *In artem analyticam isagoge* (*Introduction to the Analytic Art*). The English mathematician Thomas Harriot preferred the protocol in lowercase letters and used it to tremendous effect. This protocol is no longer in common use.

whole numbers: The numbers 0, 1, 2, 3, ...; in other words, the counting numbers and 0 or the nonnegative integers.

Notes

INTRODUCTION

1. Letter from Frederick to Voltaire, May 16, 1749: Aldington, *Letters of Voltaire and Frederick the Great*, pp. 195–196.

2. Dean, "What Price Algebra."

3. Algebra texts for use in colleges and specialized schools started to appear after 1660 both in France and at the British Royal Navy school of Christ's Hospital— and elsewhere later. Ellerton, Kanbir, and Clements, "Historical Perspectives on the Purposes of School Algebra"; Ponte and Guimarães, "Notes for a History of the Teaching of Algebra."

4. Paraphrase of Dewey's words: Dewey, *Art as Experience*, pp. 36–39; Wikipedia, "Art as Experience."

5. Keyser, "The Humanization of the Teaching of Mathematics."

CHAPTER 1: NUMERICAL SYMPHONIES

1. Sfard, "On Two Metaphors for Learning and the Dangers of Choosing Just One."

2. Fauvel and Van Maanen, *History in Mathematics Education: The ICMI Study*, p. 36.

3. Pinker, "College Makeover: The Matrix, Revisited."

4. Libin, "Symphony."

5. Libin, "Symphony"; Horton, *The Cambridge Companion to the Symphony*, p. 4.

6. Stanley, *The Cambridge Companion to Beethoven*, p. 13; Saccenti, Smilde, and Saris, "Beethoven's Deafness and His Three Styles."

7. Crowther et al., "Mapping Tree Density at a Global Scale."

8. MLB.com, "Batting Average: 2011"; MLB.com, "Batting Average: 1941."

9. Of note, in his early papers on relativity from 1905 to 1907, Einstein used the uppercase letter V to stand for the velocity/speed of light. It was only in 1907 that we see an abrupt shift in his notation to c. Webb, *Clash of Symbols*, pp. 140–141.

CHAPTER 2: ART OF MANEUVER

1. Holmes, *The Oxford Companion to Military History*, p. 541.

2. United States Marine Corps, *Warfighting*, p. 100.

3. "Maneuver," *Merriam-Webster Dictionary*.

4. The great German mathematician Carl Friedrich Gauss is reputed to have used this maneuver to add the numbers from 1 to 100 when he was an elementary school student. For a more detailed discussion of this common tale, see Hayes, "Gauss's Day of Reckoning."

5. Viète, *The Analytic Art*, p. 32.

6. Cardano, *The Rules of Algebra (Ars Magna)*, p. 8.

7. Abdul-Jabbar and Knobler, *Giant Steps*, p. 146.

8. Weeks, *The Discovery of the Elements*, p. 2. The original passage in German appeared in Winkler's 1897 article, "Ueber die Entdeckung neuer Elemente im Verlaufe der letzten fünfundzwanzig Jahre und damit zusammenhängendende Fragen." [On the discovery of new elements in the course of the last twenty-five years and related questions. (Google Translate German to English)] Shakespeare's quote is from the play *As You Like It* (ca. 1599): "All the world's a stage, And all the men and women merely players."

CHAPTER 3: NUMERICAL FORENSICS

1. In standard graphical (or Cartesian coordinate) language, the ensemble of values (or what we have called the cloud of values) for the simplest types of variable expressions, when sorted a certain way, take the shape of geometrical paths (often called curves). For example, when the input values of x are sorted in ascending order ($\ldots$, -2, -1, 0, 1, 2, $\ldots$), the path that the variable expression $3x + 7$ takes on is in the shape of a line in the x-y Cartesian plane. (Cartesian coordinate systems are not covered in this book.)

2. Recorde, *The Whetstone of Witte*, p. 222.

3. Heeffer, "Learning Concepts through the History of Mathematics."

4. Saunderson, *Elements of Algebra in Ten Books*, p. 94; Day, *An Introduction to Algebra (Colleges)*, p. 78; Euler, *Elements of Algebra*, p. 188.

5. "Bank," *Merriam-Webster Dictionary*.

6. "Table," *Merriam-Webster Dictionary*.

7. The way this rental works is that the cost is $20 per day plus a $30 one-time base fee. Letting x be the number of days gives $20x$ for the per-day amount plus the $30 base fee, which yields the expression for the total cost of renting for x days as $20x + 30$ dollars.

8. A slightly more appropriate way to deal with this would be to phrase it as the inequality $20x + 30 \leq 900$. Though not discussed in this book, solving linear inequalities (which also involve isolating the x) is often covered in elementary algebra courses.

CHAPTER 4: CONVERGING STREAMS AND EMERGING THEMES

1. Kline, *Mathematics and the Physical World*, p. ix.

2. Boyer, *History of Analytic Geometry*, p. 59; Mahoney, "The Beginnings of Algebraic Thought in the Seventeenth Century."

3. The thirteenth-century mathematician Jordanus Nemorarius (Jordanus de Nemore), an important mathematician in the history of algebra in Europe, did some early work with the notion of parameters. Unfortunately, it appears that the results of his use of parameters (particularly in the case of quadratic equations) didn't germinate and was mostly overlooked by his successors, leaving the way open for Viète's more intensified rediscovery and exploitation to finally take root 300+ years later. Boyer and Merzbach, *A History of Mathematics*, pp. 257–259.

4. Katz and Parshall, *Taming the Unknown*, p. 158.

5. Unfortunately, Viète was highly and purposefully dismissive of the contributions by medieval Islamic mathematicians, claiming that he was rediscovering algebra as formulated by the Greeks and desecrated by the "barbarians." The historical record shows that Viète was incorrect and unfair in this assessment and that the work of medieval Islamic mathematicians (such as Al-Khwarizmi, Abu Kamil, Al-Karaji, and others) as well as pre-Renaissance European mathematicians (such as Jordanus, Leonardo de Pisa, and other abbaco masters) was substantial—with some of it critically contributing to the discoveries obtained by sixteenth-century mathematicians, including Viète himself.

CHAPTER 5: THE RULE OF DARK POSITION

1. *Shay, jidhr, radix*: Rahman, Street, and Tahiri, *The Unity of Science in the Arabic Tradition*, pp. 93–94. *Cosa, coss*: Katz and Pershall, *Taming the Unknown*, pp. 194, 198, 203, 205, 206. The German algebraists use of the word *coss* was an adaptation of the word *cosa* meaning "thing" in Italian. *Yāvattāvat*: Plofker, *Mathematics in India*, pp. 59, 193.

2. Recorde, *The Whetstone of Witte*, p. 220.

3. Stedall, *A Discourse Concerning Algebra*, pp. 6, 38.

4. Mazur and Pesic, "On Mathematics, Imagination & the Beauty of Numbers."

5. The 180 degrees sum rule for the three angles of a triangle holds for triangles in a flat space. If the space on which the triangle sits is itself curved (such as a sphere), then other rules may apply.

6. In some cases, both the correct answer and the mistake could be positive, which might make the error harder to detect.

7. Al-Khwārizmī, *Al-Kitāb al-mukhtaṣar fī ḥisāb al-jabr wa'l-muqābala*, p. 5.

8. Al-Khwārizmī, *Al-Kitāb al-mukhtaṣar fī ḥisāb al-jabr wa'l-muqābala*, p. 5.

9. Al-Khwārizmī, *Al-Kitāb al-mukhtaṣar fī ḥisāb al-jabr wa'l-muqābala*, p. 8.

10. Whitehead, *An Introduction to Mathematics*, p. 115.

CHAPTER 6: THE GRAND PLAY

1. Dewey, *John Dewey: The Later Works 1925–1953*, Volume 10, pp. 372–373.

2. "Dewey to Talk on 'Art, Aesthetic Experience,'" *The Harvard Crimson*.

3. Dewey, *Art as Experience*, p. 253.

4. Dewey, *Art as Experience*, p. 2.

5. Dewey, *Art as Experience*, pp. 37, 39–40.

6. The Austrian-American physicist Victor Weisskopf gave this description of the joy of insight in the beginning of his 1991 book of the same name. Weisskopf, *The Joy of Insight*, p. vii.

7. Adler, *Stella Adler: The Art of Acting*, p. 85.

8. "Hold Infinity in the palm of your hand" from the William Blake poem "Auguries of Innocence."

9. Willingham, "What the NY Times Doesn't Know About Math Instruction."

10. Willingham, *Why Don't Students Like School?*, pp. 10–13, 63–64.

11. Willingham, *Why Don't Students Like School?*, pp. 63–64.

12. Tall and Thomas, "Encouraging Versatile Thinking in Algebra Using the Computer."

13. Doxiadis, "Embedding Mathematics in the Soul," p. 22.

14. I am not advocating for completely avoiding more complicated numbers and maneuvers in problems, but am simply suggesting that if such situations are allowed to dominate the discussion, especially initially, students may more easily lose sight of the algebraic tasks at hand.

15. National Education Association of the United States, *Report of the Committee on Secondary School Studies*, p. 108.

16. Shapin and Schaffer, *Leviathan and the Air-Pump*.

17. Paret, *Clausewitz and the State*, p. 185. Paret cites, as a source for Clausewitz's statement, a passage from the 1878 biography *Leben des Generals Clausewitz und der Frau Marie von Clausewitz* by Karl Schwartz.

CHAPTER 7: ALGEBRAIC AWARENESS

1. Grendler, *Renaissance Education Between Religion and Politics*, p. 6; Cohen and Cohen, *Daily Life in Renaissance Italy*, p. 139; Kleinhenz, *Medieval Italy*, pp. 354–355.

2. Cohen, "Numeracy in Nineteenth-Century America," p. 52.

3. Harvard was not the leader in this endeavor. Other colonial colleges such as Yale (through President Thomas Clap's efforts) had decades earlier preceded Harvard in this and other mathematical requirements. Nathaniel Hammond's algebra text, *The Elements of Algebra in a New and Easy Method*, was used during part of Clap's tenure and exhibited remarkable exposition for the era, owing to an extended introduction that gave a historical overview of the subject as well as its comparatively gentle dive into the subject. Hammond also provided copious and coordinated examples throughout the book. Cohen, *A Calculating People*, p. 123; Ellerton and Clements, *Rewriting the History of School Mathematics in North America 1607–1861*, p. 78.

4. Littlefield, *Early Schools and School-Books of New England*, pp. 179–182.

5. *Reports on the Course of Instruction in Yale College*, pp. 32–33, 53.

6. Pike, *A New and Complete System of Arithmetic*, p. 473; Simons, *Bibliography of Early American Textbooks on Algebra*, p. 8.

7. Simons, *Bibliography of Early American Textbooks on Algebra*, p. 2.

8. Simons, *Bibliography of Early American Textbooks on Algebra*, p. 3. In 1819, Jeremiah Day published a revised, mildly abbreviated edition of his 1814 algebra text addressed to colleges (*An Introduction to Algebra Being the First Part in a Course in Mathematics: Adapted to the Method of Instruction in the American Colleges*) to now include high schools and academies (*An Introduction to Algebra Being the First Part in a Course in Mathematics: Adapted to the Method of Instruction in the Higher Schools and Academies in the United States*).

9. Ravitch, *Left Back*, pp. 48, 49; Angus and Mirel, *The Failed Promise of the American High School, 1890–1995*, p. 7.

10. National Education Association of the United States, *Session of the Year 1891*, pp. 620–631, 829–830; Mackenzie, "The Report of the Committee of Ten"; Angus and Mirel, *The Failed Promise of the American High School, 1890–1995*, pp. 6–7, 10.

11. Ravitch, *Left Back*, p. 41.

12. National Education Association of the United States, *Report of the Committee on Secondary School Studies*, pp. 3, 4; National Education Association of the United States, *Session of the Year 1892*, pp. 31, 754.

13. National Education Association of the United States, *Report of the Committee on Secondary School Studies*, p. 5.

14. National Education Association of the United States, *Report of the Committee on Secondary School Studies*, p. 11.

15. National Education Association of the United States, *Report of the Committee on Secondary School Studies*, pp. 7, 8.

16. National Education Association of the United States, *Report of the Committee on Secondary School Studies*, pp. 7, 8.

17. National Education Association of the United States, *Report of the Committee on Secondary School Studies*, pp. 11–12.

18. Ravitch, *Left Back*, pp. 48–50; Angus and Mirel, *The Failed Promise of the American High School, 1890–1995*, pp. 8–17.

19. Reeve, "Attacks on Mathematics and How to Meet Them"; Bestor, *Educational Wastelands*; Cairns, "Mathematics and the Educational Octopus"; Mirel and Angus, "High Standards for All?"; Angus and Mirel, *The Failed Promise of the American High School, 1890–1995*, pp. 167–176.

20. "The Letter of Admiral Nimitz," *The American Mathematical Monthly*; Schorling, "Has There Been a Pearl Harbor in Public Education?"

21. "Harvard's Elective System," *The Harvard Crimson*; Eliot, "Shortening and Enriching the Grammar-School Course."

22. Cajori, *The Teaching and History of Mathematics in the United States*, p. 100.

23. Committee of Ten member James C. Mackenzie, headmaster of the Lawrenceville School in New Jersey, hailed from Scotland and was from a modest background. He received almost no formal education prior to his 18th year. Another committee member, Oscar D. Robinson, was a civil war veteran and a nontraditional student in that, while attending a private academy, he completed his studies there at the age of 23 in July 1862. After this, instead of going directly to college, he entered the Union Army as a private in the Ninth New Hampshire volunteer infantry, reportedly seeing action in the major engagements at Antietam, Fredericksburg, Vicksburg, Cold Harbor, and Petersburg. Of the 12 to 15 schoolmates that joined the Ninth with him, only two were left to be discharged from service at war's end. He entered Dartmouth College afterward and received his degree in 1869 at age 30. Mackenzie, "James C. Mackenzie Papers"; Robinson, "Dr. Oscar D. Robinson."

24. National Education Association of the United States, *Report of the Committee on Secondary School Studies*, p. 51.

25. National Education Association of the United States, *Report of the Committee on Secondary School Studies*, pp. 17, 51; Mackenzie, "The Report of the Committee of Ten."

26. It shouldn't be forgotten that segregation was still firmly entrenched in certain regions of America at this time, being further bolstered by the *Plessy v. Ferguson* (1896) and *Cummings v. Richmond Board of Education* (1899) Supreme Court rulings (and all that these entailed) yet to occur a little over two and five years, respectively, after the release of the Committee of Ten report.

27. Harris, "High School—Report of Principal."

28. Hall, *Adolescence*, p. 510.

29. Ravitch, *Left Back*, pp. 43–44.

30. National Education Association of the United States, *Report of the Committee on Secondary School Studies*, pp. 105, 107.

31. National Education Association of the United States, *Report of the Committee on Secondary School Studies*, p. 105. It is worth noting that, though not a member of the Committee of Ten or any of the nine subcommittees, NEA chairman Norman A. Calkins wrote an entire book on the idea of using concrete examples in instruction: Calkins, *Primary Object Lessons for Training the Senses and Developing the Faculties of Children*.

32. Simons, *Bibliography of Early American Textbooks on Algebra*; King, *An Analysis of Early Algebra Textbooks Used in the American Secondary Schools Before 1900*, pp. 176–177. It is worth noting that Angie Turner King, on the faculty at West Virginia State College, was highly influential in the education of the mathematician Katherine Johnson of NASA fame. Donoghue, "Algebra and Geometry Textbooks in Twentieth-Century America."

33. National Education Association of the United States, *Report of the Committee on Secondary School Studies*, p. 107.

34. Brown, Obourn, and Kluttz, *Offerings and Enrollments in Science and Mathematics in Public High Schools*, p. 29.

35. Baker and Jones, *Encyclopedia of Bilingualism and Bilingual Education*, p. 698.

36. Poincaré, *Science and Method*, p. 34.

37. Thurston, "Mathematical Education."

38. Whitehead, *The Aims of Education and Other Essays*, pp. 70–71.

CHAPTER 8: ALGEBRA UNCLOAKED

1. The conclusions of the problem work for whole number values of x greater than 1 through 7, but for such values, of course, the "days of the week" interpretation loses its meaning.

2. *Encyclopedia Britannica* defines an analog computer as "any of a class of devices in which continuously variable physical quantities such as electrical potential, fluid pressure, or mechanical motion are represented in a way analogous to the corresponding quantities in the problem to be solved." This setup of treating water levels to calculate course averages would qualify as an approximate type of analog computation under this definition. Gregersen, "Analog Computer."

3. It will give the exact value if the final exam averages for each class section were not rounded off (that is, if they were exactly 96, 88, and 54, respectively). It will give a relatively closer value (than the non-weighted average of 79) if any of

the three class averages were rounded off (for example, 96.4 rounded off to 96 for Section 001, 87.8 rounded off to 88 for Section 002, and so on). Try it for yourself by making up a sample of 20 final exam scores for a course with 20 students spread over three sections, where Section 001 has 5 students, Section 002 has 11 students, and Section 003 has 4 students, corresponding to the respective proportions of 25%, 55%, and 20% (as was the case when there were 100 students).

4. Descartes, *La Géométrie*, pp. 9, 33, 34, 52; Day, *An Introduction to Algebra (Colleges)*, p. 12.

5. Plofker, *Mathematics in India*, p. 193.

6. Plofker, *Mathematics in India*, pp. 193, 196.

7. MLB.com, "Slugging Percentage."

8. Late in 2020, Major League Baseball made the decision to begin including the statistics from the Negro Leagues from 1920 to 1948. While research is still ongoing, the highest slugging percentage in a season will probably end up being either Josh Gibson's 1937 season (0.974) or Mule Suttles 1926 season (0.877). Baseball Reference, "Single-Season Leaders & Records for Slugging %."

9. Ortega y Gasset, *The Dehumanization of Art*, p. 33.

10. If we had five numbers represented by x, y, z, u, and v, then we would calculate the traditional average by adding them up and dividing by 5 like so: $\dfrac{x+y+z+u+v}{5}$. However, with a bit of maneuvering, we can rewrite this as

$$\frac{x+y+z+u+v}{5}=\frac{x}{5}+\frac{y}{5}+\frac{z}{5}+\frac{u}{5}+\frac{v}{5}=\frac{1}{5}x+\frac{1}{5}y+\frac{1}{5}z+\frac{1}{5}u+\frac{1}{5}v.$$

Because $\dfrac{1}{5}$ corresponds to 0.20, we could rewrite the rightmost expression as $0.20x + 0.20y + 0.20z + 0.20u + 0.20v$. In quantitative cocktail language, this can be interpreted as five categories represented by x, y, z, u, and v each contributing 20% to the final result.

11. Morse and Brooks, "How U.S. News Calculated the 2021 Best Colleges Rankings."

12. The inclusion of the rankings here is in no way an endorsement of college rankings in general. The author is fully aware of the challenges associated with such ranking algorithms and the diversity of opinion on this highly influential issue.

CHAPTER 9: ALGEBRAIC FLIGHTS: MECHANISM AND CLASSIFICATION

1. This is illustrated even in the case where the speed is unknown at first. If we know the distance and time of travel, we can use the formula $d = st$ to solve for the

speed s. Once we know the value of s, however, we can insert it into the formula and generally treat it as a constant from then on for other problems involving that specific scenario.

2. Colonel Tina Hartley and V. Frederick Rickey, of the United States Military Academy's Mathematical Sciences Department, stated in a 2011 presentation that the earliest example they could find of m being used as the slope is from the book by John Hymers: *A Treatise on Conic Sections and the Application of Algebra to Geometry* (1837). Hartley and Rickey, "Why Do We Use 'm' for Slope?"

3. "Radioactivity," *Merriam-Webster Dictionary*.

4. Making estimates such as this one will only yield exactly correct values for the simplest variations, often called linear variations. The situation involving radioactive half-lives varies to a different tune called exponential variation, and linear interpolation can only give an approximate answer.

5. As another example, consider $t = 5$ years. Replacing t by 5 in the formula yields

$$A = 512\left(\frac{1}{2}\right)^5 = 512\left(\frac{1}{2}\right)\left(\frac{1}{2}\right)\left(\frac{1}{2}\right)\left(\frac{1}{2}\right)\left(\frac{1}{2}\right) = 512\left(\frac{1}{32}\right) = \left(\frac{512}{1}\right)\left(\frac{1}{32}\right) = \frac{512}{32}$$
$$= 16 \text{ ounces.}$$

This matches the amount given in the table for 5 years. For $t = 0$, we have $\left(\frac{1}{2}\right)^0$, which equals one. This is not an intuitive result, but it can be demonstrated algebraically why the zeroth power should take on this value. Though not discussed here, the reasoning can be found in many textbooks on algebra.

6. Consider $t = 3$ years for the formula $A = 816\left(\frac{1}{2}\right)^t$. This yields

$$A = 816\left(\frac{1}{2}\right)^3 = 816\left(\frac{1}{2}\right)\left(\frac{1}{2}\right)\left(\frac{1}{2}\right) = 816\left(\frac{1}{8}\right) = \frac{816}{8} = 102 \text{ ounces.}$$

This matches the predicted value in the table.

7. Check that when $t = 15$ years for the formula $A = 816\left(\frac{1}{2}\right)^{\frac{t}{3}}$, we obtain

$$A = 816\left(\frac{1}{2}\right)^{\frac{15}{3}} = 816\left(\frac{1}{2}\right)^5 = 816\left(\frac{1}{2}\right)\left(\frac{1}{2}\right)\left(\frac{1}{2}\right)\left(\frac{1}{2}\right)\left(\frac{1}{2}\right) = 816\left(\frac{1}{32}\right) = \frac{816}{32}$$
$$= 25.5 \text{ ounces.}$$

This matches the predicted value in the table.

8. Splinter, *Illustrated Encyclopedia of Applied and Engineering Physics*, p. 201.

9. Splinter, *Illustrated Encyclopedia of Applied and Engineering Physics*, p. 201.

10. Here is the calculation of the midpoint value for the expanded IM-5730 table to estimate the time for a sample to get to 1.64 ounces: For the original table with six entries, 1.64 ounces is between 1 ounce and 2 ounces. The halfway point between the years for each of these values in the table is

$$\frac{17190 \text{ years} + 11460 \text{ years}}{2} = 14325 \text{ years}.$$

This value differs from the correct value by 1225 years (14325 – 13100 = 1225).

For the expanded table with 12 entries, 1.64 ounces is now between 1.5 ounces and 2 ounces. The halfway point between the years for each of these values is

$$\frac{13838.2 \text{ years} + 11460 \text{ years}}{2} = 12649.1 \text{ years}.$$

This value differs from the correct value by approximately 451 years (13100 – 12649.1 = 450.9).

11. Though scholasticism got a bad rap from many scientists, philosophers, and mathematicians of the early scientific era, it represented a revolution in its own right from thinking in earlier medieval periods. Some of the biggest achievements produced by its practitioners included the grand effort to place theology on a foundation of logic (primarily as developed by Aristotle) and reason along with the development and systematic use of much more efficient ways to organize and synthesize large, diverse bodies of information as an aid in that endeavor. Some of these techniques were to prove exceedingly fruitful when adapted to a different platform, that of experimentation and mathematics. Well-known scholastics included St. Anselm, Peter Abelard, Peter Lombard, St. Albertus Magnus, Roger Bacon, St. Thomas Aquinas, and William of Ockham. Longwell, "The Significance of Scholasticism."

CHAPTER 10: ALGEBRAIC FLIGHTS: INDETERMINACY AND CURIOSITY

1. O'Connor and Robertson, "Diophantus of Alexandria."

2. Schappacher, *Diophantus of Alexandria*.

3. Heath, *Diophantus of Alexandria*, p. 124.

4. The way Diophantine equations are usually introduced requires that the coefficients or parameters be integers. This is straightforward to do here by simply multiplying both sides of the equation $0.20x + 0.60y + 0.20z = 80$ by 10. Remember, if we multiply both sides of an equation by the same nonzero number, we change

the form of the equation but not the solutions. Doing so here unveils the equation $2x + 6y + 2z = 800$ (which transforms the coefficients 0.20, 0.60, and 0.20 to the new coefficients 2, 6, and 2 and turns 80 into 800). This new equation will have the same integer solutions as the course average equation, as long as we stipulate that the blended average of the three scores (homework, tests, and final exam) must add to exactly 80 and not be rounded off to 80.

5. For a demonstration of why positive integer solutions exist only for $x + y + z = 12$ (months), see Hayes et al., "Linear Diophantine Equations."

6. The number 1 is the lone exception because it is equal to itself squared.

7. The base of the exponent can be thought of as being the generator of the perfect square; that is, the number 7 can be thought of as the generator of 49 because 7×7 generates 49, or 13 can be thought of as the generator of 169 because $13 \times 13 = 169$. Medieval mathematicians often called this generator the *root*, and through their viewpoint naturally extended the name *root* to include values that satisfied any of their second-order equations. Here, we can call 7 a root of the perfect square 49 (or 13 a root of the perfect square 169).

However, 7 can serve as the generator of other values as well. For instance, if we multiply 7 three times, we get 343 ($7 \times 7 \times 7 = 343$), or if we multiply 7 four times, we get 2401 ($7 \times 7 \times 7 \times 7 = 2401$). This means that 7 can also be seen to be a root of 343 and 2401. Using the same term can get confusing, as the way that 7 generates 49 is different from the way it generates 343, which in turn is different from the way it generates 2401, and so on.

To distinguish the various types of generations of a number such as 7, mathematicians give names to the various types of roots: square root (for roots that generate values from being multiplied twice, or squared), cube root (for roots that generate values from being multiplied three times, or cubed), and fourth root (for roots that generate values from being multiplied four times). Thus, 7 is the square root of 49 because it generates 49 by being multiplied twice ($49 = 7 \times 7$ or 7^2). Similarly, 7 is the cube or third root of 343 because it generates 343 by being multiplied three times ($343 = 7 \times 7 \times 7$ or 7^3), and 7 is the fourth root of 2401 because it generates 2401 by being multiplied four times ($2401 = 7 \times 7 \times 7 \times 7$ or 7^4). In a similar fashion, 13 is the square root of 169, the cube root of 2197, and the fourth root of 28,561.

It is important to note, however, that in describing the solutions of a general polynomial equation, the generic term *root* is still in use.

8. The quantities represented by m and n are also sometimes called parameters by mathematicians, but they are not parameters in the exact sense that we have used the term in this book (as scenario variables—fixed within a scenario alongside other quantities that are varying within the scenario).

Sometimes varying quantities, such as the three sides of numerous types of right triangles, can be constructed from other more hidden quantities. But the exact nature of these hidden relationships is sometimes not immediately obvious from their visibly expressed relationships in the form of the Pythagorean Theorem ($x^2 + y^2 = z^2$). When mathematicians unearth these hidden relationships to the other quantities, they often introduce a new set of variables to illustrate and represent the connections. This new set of quantities is often also called a *set of parameters*. This is what has happened here with using m and n.

Clearly, the sides of the right triangles, represented by x, y, and z, are related to each other through the Pythagorean Theorem ($x^2 + y^2 = z^2$). But they are also interrelated in an interesting way to another set of quantities—represented by m and n—that allows us to easily construct integer-valued sides. The splitting of the three unknowns x, y, and z into the respective relationships $m^2 - n^2$, $2mn$, and $m^2 + n^2$ gives form to this construction through allowing us to simply choose positive integer values for m and n. These chosen integer values can then, in turn, automatically yield positive integer results for the sides x, y, and z—thus giving us an abundance of the positive integer sides that we sought (as long as m is larger than n).

9. The Babylonians used a base 60 system, called sexagesimal, to represent numbers that, though equivalent in value to our base 10 values such as 45 and 75, have a dramatically different look on Plimpton 322. Neugebauer and Sachs, *Mathematical Cuneiform Texts*, Plate 25.

See a brief discussion of sexagesimal numerals in my earlier book: Williams, *How Math Works*, pp. 212–213.

10. Neugebauer and Sachs, *Mathematical Cuneiform Texts*, pp. 38–41.

11. Eleanor Robson, "Words and Pictures: New Light on Plimpton 322."

12. Mathematical Association of America, *Paul R. Halmos–Lester R. Ford Awards*.

13. Mansfield and Wildberger, "Plimpton 322 Is Babylonian Exact Sexagesimal Trigonometry"; Moss, "Was Geometry Invented by Bureaucrats and Not a Greek Genius?"; Engelking, "Can This Ancient Babylonian Tablet Improve Modern Math?"; Lamb, "Don't Fall for Babylonian Trigonometry Hype."

14. Høyrup, "Pythagorean 'Rule' and 'Theorem.'"

15. Clapham and Nicholson, *The Concise Oxford Dictionary of Mathematics*, p. 170.

16. Bell, *Men of Mathematics*, p. 419.

17. Matiyasevich, *Hilbert's Tenth Problem*, pp. 2, 4.

18. François Viète, *The Analytic Art*, pp. 67–68.

CHAPTER 11: A KALEIDOSCOPE OF INGREDIENTS

1. David Pimm, *Symbols and Meanings in School Mathematics*, pp. 88, 122.
2. Roscoe, *The Life and Experiences of Sir Henry Roscoe*, p. 81.
3. Sagan, *Cosmos*, p. 93.
4. MIT Spectroscopy, "Nobel Prizes."
5. Butcher, *Tour of the Electromagnetic Spectrum*, p. 2.
6. Chomsky, *For Reasons of State*, p. 402.
7. Other important properties that are often affiliated with closure, such as associativity, inverses, and identities, were not pointed out here but are present in real numbers and by consequence some were leveraged in the algebraic processes used.

CHAPTER 12: GRAND CONFLUENCES

1. *Reports on the Course of Instruction in Yale College*, p. 14.
2. Yale University, "Yale University. University Catalogue, 1828."
3. Winterer, *The Culture of Classicism*, pp. 32–34, 60.
4. *Reports on the Course of Instruction in Yale College*, pp. 11, 14–15.
5. Pereltsvaig, *Languages of the World*, p. 11.
6. Descartes, *La Géométrie*, pp. 9, 33, 34, 52.
7. Fontenelle, *Eloges des academiciens de l'Academie royale des sciences*, Preface.
8. Calvino actually said, "The word connects the visible trace with the invisible thing, the absent thing, the thing that is desired or feared like a frail emergency bridge flung over an abyss." Calvino, *Six Memos for the Next Millennium*, p. 77.
9. Eisner, *The Arts and the Creation of Mind*, p. 11.
10. Thurston, Foreword to *Teichmuller Theory and Applications to Geometry, Topology, and Dynamics*.
11. Willingham, "What the NY Times Doesn't Know About Math Instruction."
12. Dewey, *Art as Experience*, pp. 36–39.

Bibliography

Abdul-Jabbar, Kareem, and Peter Knobler. *Giant Steps: The Autobiography of Kareem Abdul-Jabbar*. New York: Bantam Books, 1984.

Adler, Stella. *Stella Adler: The Art of Acting*. Edited by Howard Kissel. New York: Applause Books, 2000.

Aldington, Richard, translator. *Letters of Voltaire and Frederick the Great*. New York: Brentano's, 1927.

Al-Khwārizmī, Muḥammad ibn Mūsā. *Al-Kitāb al-mukhtaṣar fī ḥisāb al-jabr wa'l-muqābala* (*The Compendious Book on Calculation by Completion and Balancing*) (ca. 830). In *The Algebra of Mohammed Ben Musa*, edited and translated by Frederic Rosen. London: J. Murray, 1831.

Angus, David L., and Jeffrey E. Mirel. *The Failed Promise of the American High School, 1890–1995*. New York: Teachers College Press, 1999.

Baker, Colin, and Sylvia Prys Jones, editors. *Encyclopedia of Bilingualism and Bilingual Education*. Clevedon: Multilingual Matters, 1998.

"Bank." *Merriam-Webster Dictionary*. Accessed May 30, 2021. www.merriam-webster.com/dictionary/bank.

Baseball Reference. "Single-Season Leaders & Records for Slugging %." Accessed October 8, 2021. www.baseball-reference.com/leaders/slugging_perc_season.shtml.

Bell, E. T. *Men of Mathematics: The Lives and Achievements of the Great Mathematicians from Zeno to Poincaré*. Melbourne: Penguin Books, 1937.

Bestor, Arthur E. *Educational Wastelands: The Retreat from Learning in Our Public Schools*. Urbana: University of Illinois Press, 1953.

Blake, William. "Auguries of Innocence." In *The Complete Poetry and Prose of William Blake*, edited by David V. Erdman, p. 490. Berkeley: University of California Press, 2008.

Boyer, Carl. *History of Analytic Geometry*. Mineola: Dover Publications, Inc., 2004.

Boyer, Carl, and Uta C. Merzbach. *A History of Mathematics*, Third Edition. Hoboken: John Wiley & Sons, Inc., 2011.

Brown, Kenneth E., Ellsworth S. Obourn, and Marguerite Kluttz. *Offerings and Enrollments in Science and Mathematics in Public High Schools*. Washington, DC: Government Printing Office, 1958.

Butcher, Ginger. *Tour of the Electromagnetic Spectrum*. Washington, DC: National Aeronautics and Space Administration, 2010.

Cairns, Stewart Scott. "Mathematics and the Educational Octopus." *The Scientific Monthly* 76, no. 4 (April 1953), pp. 231–240.

Cajori, Florian. *The Teaching and History of Mathematics in the United States*. Washington, DC: Government Printing Office, 1890.

Calkins, N. A. *Primary Object Lessons for Training the Senses and Developing the Faculties of Children: A Manual of Elementary Instruction for Parents and Teachers*. New York: Harpers & Brothers, 1870.

Calvino, Italo. *Six Memos for the Next Millennium*. Translated by Patrick Creagh. Cambridge, MA: Harvard University Press, 1988.

Cardano, Girolamo. *The Rules of Algebra (Ars Magna)* (1545). Edited and Translated by T. Richard Witmer. Mineola: Dover Publications Inc., 1968.

Chomsky, Noam. *For Reasons of State*. New York: Pantheon Books, 1973.

Clapham, Christopher, and James Nicholson. *The Concise Oxford Dictionary of Mathematics*, Fourth Edition. Oxford: Oxford University Press, 2009.

Cohen, Elizabeth S., and Thomas V. Cohen. *Daily Life in Renaissance Italy*. Westport: Greenwood Press, 2001.

Cohen, Patricia Cline. *A Calculating People: The Spread of Numeracy in Early America*. New York: Routledge, 1999.

Cohen, Patricia Cline. "Numeracy in Nineteenth-Century America." In *A History of School Mathematics*, edited by George M. A. Stanic and Jeremy Kilpatrick, pp. 43–71. Reston, VA: National Council of Teachers of Mathematics, 2003.

Crowther, T. W., H. B. Glick, K. R. Covey, C. Bettigole, D. S. Maynard, S. M. Thomas, J. R. Smith, et al. "Mapping Tree Density at a Global Scale." *Nature* 525 (2015), pp. 201–205.

Day, Jeremiah. *An Introduction to Algebra Being the First Part in a Course in Mathematics: Adapted to the Method of Instruction in the American Colleges*. New Haven: Howe & Deforest, 1814.

Day, Jeremiah. *An Introduction to Algebra Being the First Part in a Course in Mathematics: Adapted to the Method of Instruction in the Higher Schools and Academies in the United States*. New Haven: Howe & Spalding, 1819.

Dean, Arthur. "What Price Algebra." Your Boy and Your Girl. *The Olean Evening Times*, March 27, 1930.

Descartes, René. *La Géométrie* (1637). In *The Geometry of René Descartes: Translated from the French and Latin, with a Facsimile of the First Edition 1637*, translated by David Eugene Smith and Marcia L. Latham. Chicago: The Open Court Publishing Company, 1925.

Devlin, Keith. "The Math Myth That Permeates 'The Math Myth.'" *Devlin's Angle*, March 1, 2016. Accessed May 30, 2021. http://devlinsangle.blogspot.com/2016/03/the-math-myth-that-permeates-math-myth.html.

Dewey, John. *Art as Experience* (1934), Trade Paperback Edition. New York: Berkley Publishing Group, 2005.

Dewey, John. *John Dewey: The Later Works, 1925–1953*, Volume 10, *1934: Art as Experience*. Edited by Jo Ann Boydston. Carbondale/Edwardsville: Southern Illinois University Press, 2008.

"Dewey to Talk on 'Art, Aesthetic Experience': Inglis Lecturer Presents a Series of Ten Addresses." *The Harvard Crimson*. February 24, 1931.

Donoghue, Eileen F. "Algebra and Geometry Textbooks in Twentieth-Century America." In *A History of School Mathematics*, edited by George M. A. Stanic and Jeremy Kilpatrick, pp. 329–398. Reston, VA: National Council of Teachers of Mathematics, 2003.

Doxiadis, Apostolos. "Embedding Mathematics in the Soul: Narrative as a Force in Mathematics Education." Transcript of the Opening Address to the Third Mediterranean Conference on Mathematics Education, Athens, Greece, January 3–5, 2003.

Eisner, Elliot W. *The Arts and the Creation of Mind*. New Haven: Yale University Press, 2002.

Eliot, Charles. "Shortening and Enriching the Grammar-School Course." *School and College: Devoted to Secondary and Higher Education* 1, no. 3 (1892), pp. 153–164.

Ellerton, Nerida F., and M. A. (Ken) Clements. *Rewriting the History of School Mathematics in North America 1607–1861: The Central Role of Cyphering Books*. Dordrecht: Springer, 2012.

Ellerton, Nerida F., Sinan Kanbir, and M. A. Clements. "Historical Perspectives on the Purposes of School Algebra." In *Proceedings of the 40th Annual Meeting of the Mathematics Education Research Group of Australasia*, edited by Ann Dowton, Sharyn Livy, and Jennifer Hall, pp. 221–228. Adelaide: Mathematics Education Research Group of Australasia, 2017.

Engelking, Carl. "Can This Ancient Babylonian Tablet Improve Modern Math?" *Discover Magazine D-Brief*. August 24, 2017. Accessed July 4, 2021. www.discovermagazine.com/planet-earth/can-this-ancient-babylonian-tablet-improve-modern-math.

Euler, Leonard. *Elements of Algebra*, Fourth Edition. Translated by Rev. John Hewlett. London: Longman, Rees, Orme and Co., 1828.

Fauvel, John, and Jan Van Maanen, editors. *History in Mathematics Education: The ICMI Study*. Dordrecht: Kluwer Academic Publishers, 2000.

Fontenelle, Bernard le Bovier de. *Eloges des academiciens de l'Academie royale des sciences*, Volume 1. The Hague: Isaac van der Kloot, 1731.

Gregersen, Erik. "Analog Computer." *Encyclopedia Britannica*. Accessed May 21, 2021. www.britannica.com/technology/analog-computer.

Grendler, Paul F. *Renaissance Education Between Religion and Politics*. Aldershot: Ashgate Publishing Limited, 2006.

Hall, G. Stanley. *Adolescence: Its Psychology and Its Relations to Physiology, Anthropology, Sociology, Sex, Crime, Religion and Education*, Volume 2. New York: D. Appleton and Company, 1904.

Harris, W. T. "High School—Report of Principal." In *Nineteenth Annual Report of the Board of Directors of the St. Louis Public Schools, for the Year Ending August 1, 1873*, pp. 54–104. St. Louis: Democrat Litho. and Printing Co., 1874.

Hartley, Tina, and V. Frederick Rickey. "Why Do We Use 'm' for Slope?" Presentation at the History and Pedagogy of Mathematics Section Meeting, American University, Washington, DC, March 12–13, 2011.

"Harvard's Elective System: Comments of the New York Times." *The Harvard Crimson*. May 3, 1883.

Hayes, Andy, John Ashley Capellan, Khang Nguyen Thanh, Borut Levart, Ankit Nigam, Worranat Pakornrat, Margaret Zheng, et. al. "Linear Diophantine Equations." *Brilliant.org*. Accessed July 1, 2021. https://brilliant.org/wiki/linear-diophantine-equations-one-equation/.

Hayes, Brian. "Gauss's Day of Reckoning." *American Scientist* 94, no. 3 (2006), pp. 200–205.

Heath, Thomas L. *Diophantus of Alexandria: A Study in the History of Greek Algebra*, Second Edition. Cambridge: Cambridge University Press, 1910.

Heeffer, Albrecht. "Learning Concepts through the History of Mathematics: The Case of Symbolic Algebra." In *Philosophical Dimensions in Mathematics Education*, edited by Karen François and Jean Paul Van Bendegem, pp. 83–104. New York: Springer Science, 2007.

Holmes, Richard, editor. *The Oxford Companion to Military History*. Oxford: Oxford University Press, 2001.

Horton, Julian. *The Cambridge Companion to the Symphony*. Cambridge: Cambridge University Press, 2013.

Høyrup, J. "Pythagorean 'Rule' and 'Theorem'—Mirror of the Relation between Babylonian and Greek Mathematics." In *Babylon: Focus mesopotamischer Geschichte, Wiege früher Gelehrsamkeit, Mythos in der Moderne (CDOG 2)*, edited by J. Renger, pp. 393–407. Saarbrucken: Saarbrücker Druckerei und Verlag, 1999.

Katz, Victor J., and Karen Hunger Parshall. *Taming the Unknown: A History of Algebra from Antiquity to the Early Twentieth Century*. Princeton: Princeton University Press, 2014.

Keyser, Cassius Jackson. "The Humanization of the Teaching of Mathematics." *Science* 35, no. 904 (April 26, 1912), pp. 637–647.

King, Angie Turner. *An Analysis of Early Algebra Textbooks Used in the American Secondary Schools Before 1900.* Doctoral Dissertation. Pittsburgh: University of Pittsburgh Press, 1955.

Kleinhenz, Christopher, editor. *Medieval Italy: An Encyclopedia*, 2 volumes. New York: Routledge, 2004.

Kline, Morris. *Mathematics and the Physical World.* London: John Murray, 1959.

Lamb, Evelyn. "Don't Fall for Babylonian Trigonometry Hype: Separating Fact from Speculation in Math History." *Scientific American Blogs.* August 29, 2017. Accessed July 4, 2021. https://blogs.scientificamerican.com/roots-of-unity/dont-fall-for-babylonian-trigonometry-hype/.

"The Letter of Admiral Nimitz." *The American Mathematical Monthly* 49, no. 3 (March 1942), pp. 212–214.

Li, Jeffrey, Matt Lisle, Cynthia LaBrake, Paul McCord, David Laude, David Vanden Bout, and Albert Almanza. *UT Austin–Principles of Chemistry.* OpenStax CNX. May 3, 2018. Accessed July 11, 2021. http://cnx.org/contents/9d40d22b-d745-446e-889a-fb75847275dc@13.367.

Libin, Laurance Elliot. "Symphony." *Encyclopedia Britannica.* Accessed May 22, 2021. www.britannica.com/art/symphony-music.

Littlefield, George Emery. *Early Schools and School-Books of New England.* Boston: The Club of Odd Volumes, 1904.

Longwell, Horace Craig. "The Significance of Scholasticism." *The Philosophical Review* 37, no. 3 (May 1928), pp. 210–225.

Mackenzie, James C. "The Report of the Committee of Ten." *The School Review* 2, no. 3 (1894), pp. 146–155.

Mackenzie, James C. "James C. Mackenzie Papers." *The Lawrenceville School Stephan Archives*, DC033. Accessed June 13, 2021. http://archivesspace.lawrenceville.org/repositories/2/resources/34.

Mahoney, Michael S. "The Beginnings of Algebraic Thought in the Seventeenth Century." In *Descartes: Philosophy, Mathematics and Physics*, edited by S. Gaukroger, Chapter 5. Sussex: The Harvester Press, 1980.

"Maneuver." *Merriam-Webster Dictionary.* Accessed May 30, 2021. www.merriam-webster.com/dictionary/maneuver.

Mansfield, Daniel F., and N. J. Wildberger. "Plimpton 322 Is Babylonian Exact Sexagesimal Trigonometry." *Historia Mathematica* 44, no. 4 (November 2017), pp. 395–419.

Mathematical Association of America. *Paul R. Halmos–Lester R. Ford Awards.* Accessed April 4, 2018. www.maa.org/programs/maa-awards/writing-awards/paul-halmos-lester-ford-awards.

Matiyasevich, Yuri. *Hilbert's Tenth Problem*. Cambridge, MA: The MIT Press, 1996.

Mazur, Barry, and Peter Pesic. "On Mathematics, Imagination & the Beauty of Numbers." *Daedalus* 134, no. 2 (Spring 2005), pp. 124–130.

Mirel, Jefffrey, and David Angus. "High Standards for All? The Struggle for Equality in the American High School Curriculum, 1890–1990." *American Educator* 18, no. 2 (Summer 1994), pp. 4–9, 40–42.

MIT Spectroscopy. "Nobel Prizes." *History*. Accessed April 2021. http://web.mit.edu/spectroscopy/history/nobel.html.

MLB.com. "Batting Average: 1941." Accessed May 30, 2021. www.mlb.com/stats/batting-average/1941.

MLB.com. "Batting Average: 2011." Accessed May 30, 2021. www.mlb.com/stats/batting-average/2011.

MLB.com. "Slugging Percentage." Accessed May 30, 2021. http://m.mlb.com/glossary/standard-stats/slugging-percentage.

Morse, Robert, and Eric Brooks. "How U.S. News Calculated the 2021 Best Colleges Rankings." *U.S. News & World Report*, September 13, 2020. Accessed December 30, 2020. www.usnews.com/education/best-colleges/articles/how-us-news-calculated-the-rankings.

Moss, Candida. "Was Geometry Invented by Bureaucrats and Not a Greek Genius?" *The Daily Beast*. August 15, 2021. www.thedailybeast.com/was-geometry-invented-by-bureaucrats-and-not-a-greek-genius.

National Education Association of the United States. *National Education Association, Journal of Proceedings and Addresses, Session of the Year 1891, Held at Toronto, Ontario, Canada*. New York: National Education Association, 1891.

National Education Association of the United States. *National Education Association, Journal of Proceedings and Addresses, Session of the Year 1892, Held at Saratoga Springs, NY*. New York: National Education Association, 1893.

National Education Association of the United States. *Report of the Committee on Secondary School Studies Appointed at the Meeting of the National Education Association, July 9, 1892, With the Reports of the Conferences Arranged by This Committee and Held December 28–30, 1892*. Washington, DC: Government Printing Office, 1893.

Nelson, David. "Diophantine Equation." In *The Penguin Dictionary of Mathematics*, Fourth Edition. London: Penguin Books Limited, 2008.

Neugebauer, Otto, and Abraham Joseph Sachs, editors. *Mathematical Cuneiform Texts*. New Haven: American Oriental Society & American Schools of Oriental Research, 1945.

O'Connor, J. J., and E. F. Robertson. "Diophantus of Alexandria." *MacTutor: History of Mathematics*. Last Updated February 1999. Accessed April 25, 2020. https://history.mcs.st-andrews.ac.uk/Biographies/Diophantus.html.

Ortega y Gasset, José. *The Dehumanization of Art: And Other Essays on Art, Culture and Literature*. Princeton: Princeton University Press, 1968.

Paret, Peter. *Clausewitz and the State: The Man, His Theories, and His Times*. Princeton: Princeton University Press, 1985.

Pereltsvaig, Asya. *Languages of the World: An Introduction*. Cambridge: Cambridge University Press, 2012.

Pike, Nicholas. *A New and Complete System of Arithmetic: Composed for the Use of the Citizens of the United States*, Second Edition. Revised and corrected by Ebenezer Adams. Worcester: Isaiah Thomas, 1797.

Pimm, David. *Symbols and Meanings in School Mathematics*. London: Routledge, 1995.

Pinker, Steven. "College Makeover: The Matrix, Revisited." *Slate Magazine*, November 15, 2005. https://slate.com/news-and-politics/2005/11/college-makeover.html.

Plofker, Kim. *Mathematics in India*. Princeton: Princeton University Press, 2009.

Poincaré, Henri. *Science and Method*. Translated by Francis Maitland. London: Thomas Nelson and Sons, 1914.

Ponte, João Pedro da, and Henrique Manuel Guimarães. "Notes for a History of the Teaching of Algebra." In *Handbook on the History of Mathematics Education*, edited by Alexander Karp and Gert Schubring, pp. 459–472. New York: Springer, 2014.

"Radioactivity." *Merriam-Webster Dictionary*. Accessed April 1, 2018. www.merriam-webster.com/dictionary/radioactivity.

Rahman, Shahid, Tony Street, and Hassan Tahiri, editors. *The Unity of Science in the Arabic Tradition: Science, Logic, Epistemology and Their Interactions*. Dordrecht: Springer, 2008.

Ravitch, Diane. *Left Back: A Century of Failed School Reforms*. New York: Simon & Schuster, 2000.

Recorde, Robert. *The Whetstone of Witte: whiche is the Seconde parte of Arithmetike: containyng thextraction of rootes: the cossike practise, with the rule of equation: and the woorkes of surde nombers*. London: John Kyngstone, 1557.

Reeve, W. D. "Attacks on Mathematics and How to Meet Them." In *National Council of Teachers of Mathematics Eleventh Yearbook: The Place of Mathematics in Modern Education* (1936), pp. 1–21. New York: AMS Reprint Company, 1966.

Reports on the Course of Instruction in Yale College: By a Committee of the Corporation and the Academical Faculty. New Haven: Hezekiah Howe, 1828.

Robinson, Oscar D. "Dr. Oscar D. Robinson." *American Education* XV, no. 2 (September 1911), p. 26.

Robson, Eleanor. "Words and Pictures: New Light on Plimpton 322." *The American Mathematical Monthly* 109, no. 2 (2002), pp. 105–120.

Roscoe, Henry. *The Life and Experiences of Sir Henry Roscoe*. London: Macmillan and Co., 1906.

Saccenti, Edoardo, Age K. Smilde, and Wim H. M. Saris. "Beethoven's Deafness and His Three Styles." *British Medical Journal* 343, no. 7837 (December 2011), pp. 1298–1300.

Sagan, Carl. *Cosmos*. New York: Random House, 1980.

Saunderson, Nicholas. *Elements of Algebra in Ten Books*. Cambridge: Cambridge University Press, 1740.

Schappacher, Norbert. *Diophantus of Alexandria: A Text and Its History*. Published online, 2005. Accessed November 23, 2020. http://irma.math.unistra.fr/~schappa/NSch/Publications_files/1998cBis_Dioph.pdf.

Schorling, Raleigh. "Has There Been a Pearl Harbor in Public Education?" *The University of Michigan School of Education Bulletin* 13 (1941–1942), pp. 123–126.

Sfard, Anna. "On Two Metaphors for Learning and the Dangers of Choosing Just One." *Educational Researcher* 27, no. 2 (March 1998), pp. 4–13.

Shapin, Steven, and Simon Schaffer. *Leviathan and the Air-Pump: Hobbes, Boyle, and the Experimental Life: Including a Translation of Thomas Hobbes, Dialogus Physicus de Natura Aeris*. Princeton: Princeton University Press, 1985.

Simons, Lao Genevra. *Bibliography of Early American Textbooks on Algebra: Published in the Colonies and the United States through 1850, Together with a Characterization of the First Edition of Each Work*. New York: Scripta Mathematica, 1936.

Splinter, Robert. *Illustrated Encyclopedia of Applied and Engineering Physics*, Volume III, *P–Z*. Boca Raton: CRC Press, 2017.

Stanley, Glenn. *The Cambridge Companion to Beethoven*. New York: Cambridge University Press, 2008.

Stedall, Jacqueline A. *A Discourse Concerning Algebra: English Algebra to 1685*. Oxford: Oxford University Press, 2002.

"Table." *Merriam-Webster Dictionary*. Accessed May 30, 2021. www.merriam-webster.com/dictionary/table.

Tall, David, and Michael Thomas. "Encouraging Versatile Thinking in Algebra Using the Computer." *Educational Studies in Mathematics* 22, no. 2 (1991), pp. 125–147.

Thurston, William P. "Mathematical Education." *Notices of the American Mathematical Society* 37, no. 7 (1990), pp. 844–850.

Thurston, William P. Foreword to *Teichmüller Theory and Applications to Geometry, Topology, and Dynamics*, Volume 1, *Teichmüller Theory*, by John Hubbard. Ithaca: Matrix Editions, 2006.

United States Marine Corps. *Warfighting: The U.S. Marine Corps Book of Strategy: Tactics for Managing Conflict*. New York: Currency Doubleday, 1995.

Viète, François. *The Analytic Art: Nine Studies in Algebra, Geometry and Trigonometry from the Opus Restitutae Mathematicae Analyseos, seu Algebra Nova* (1590–1603). Translated by T. Richard Witmer. Kent: The Kent State Press, 1983.

Webb, Stephen. *Clash of Symbols: A Ride Through the Riches of Glyphs*. Cham: Springer International Publishing, 2018.

Weeks, Mary Elvira. *The Discovery of the Elements*, Journal of Chemical Education Seventh Edition. Revised by Henry M. Leicester. Easton, PA: Mack Printing Company, 1968.

Weisskopf, Victor. *The Joy of Insight: Passions of a Physicist*. New York: Basic Books, 1991.

Whitehead, Alfred North. *An Introduction to Mathematics*. New York: Henry Holt and Company, 1911.

Whitehead, Alfred North. *The Aims of Education*. New York: The Macmillan Company, 1929.

Whitehead, Alfred North. *The Aims of Education and Other Essays* (1929), Fourth Printing. New York: The New American Library, 1953.

Wikipedia. "Art as Experience." *Wikipedia, The Free Encyclopedia*. Accessed March 19, 2018. https://en.wikipedia.org/w/index.php?title=Art_as_Experi ence&oldid=778278335.

Williams, G. Arnell. *How Math Works: A Guide to Grade School Arithmetic for Parents and Teachers*. Lanham: Rowman and Littlefield Publishers, Inc., 2013.

Willingham, Daniel T. *Why Don't Students Like School?: A Cognitive Scientist Answers Questions about How the Mind Works and What It Means for the Classroom*. San Francisco: Jossey-Bass, 2009.

Willingham, Daniel T. "What the NY Times Doesn't Know About Math Instruction." *Daniel Willingham—Science and Education* (blog). December 9, 2013. Accessed June 12, 2021. www.danielwillingham.com/daniel-willingham-science -and-education-blog/what-the-ny-times-doesnt-know-about-math-instruction.

Winterer, Caroline. *The Culture of Classicism: Ancient Greece and Rome in American Intellectual Life, 1780–1910*. Baltimore: The Johns Hopkins University Press, 2002.

Yale University. "Yale University. University Catalogue, 1828." *Yale University Catalogue* 13 (1828), pp. 24–25.

Index

G. Arnell Williams is a professor of mathematics at San Juan College in New Mexico. He is the author of *How Math Works* and the recipient of numerous teaching awards. Williams holds degrees in physics from California State University at Long Beach and in mathematics from Yale University. He lives in Northwestern New Mexico.